THE VIRGIN AND THE MOUSETRAP

ALSO BY CHET RAYMO

In the Falcon's Claw

Honey from Stone

Soul of the Night

Biography of a Planet

The Crust of the Earth

365 Starry Nights

CHET RAYMO

THE VIRGIN

Essays in Search of

AND THE

the Soul of Science

MOUSETRAP

VIKING

VIKING
Published by the Penguin Group
Viking Penguin, a division of Penguin Books USA Inc.,
375 Hudson Street, New York, New York 10014, U.S.A.
Penguin Books Ltd, 27 Wrights Lane, London W8 5TZ, England
Penguin Books Australia Ltd, Ringwood, Victoria, Australia
Penguin Books Canada Ltd, 2801 John Street, Markham, Ontario, Canada L3R 1B4
Penguin Books (N.Z.) Ltd, 182–190 Wairau Road, Auckland 10, New Zealand

Penguin Books Ltd, Registered Offices: Harmondsworth, Middlesex, England

First published in 1991 by Viking Penguin, a division of Penguin Books USA Inc.
10 9 8 7 6 5 4 3 2 1   1 3 5 7 9 10 8 6 4 2   1 2 3 4 5 6 7 8 9 10

Some of the essays in this book are based on material from articles which appeared in *The Boston Globe* and *Sky and Telescope*.

Grateful acknowledgment is made for permission to use the following copyrighted works:
Lines from "Power" are reprinted from *The Dream of a Common Language, Poems 1974–1977*, by Adrienne Rich, by permission of the author and W. W. Norton & Company, Inc. Copyright © 1978 by W. W. Norton & Company, Inc.
Lines from a selection from *Season Songs* by Ted Hughes. By permission of Olwyn Hughes.
The Merode Altarpiece by Robert Campin is reproduced with the permission of The Metropolitan Museum of Art, New York. All rights reserved.

LIBRARY OF CONGRESS CATALOGING IN PUBLICATION DATA
Raymo, Chet.
The virgin and the mousetrap : essays in search of the soul of science / Chet Raymo.
p. cm.
ISBN 0-670-83315-0
1. Science—Methodology. 2. Science—Philosophy. 3. Knowledge, Theory of. I. Title.
Q175.R3313 1991 501—dc20 90-50762

Printed in the United States of America
Set in Primer
Designed by Jessica Shatan

For Nils Bruzelius

and Kathy Everly

The following people helped make this a better book: Nils Bruzelius and Kathy Everly of *The Boston Globe*; Lisa Kaufman, Kathryn Harrison, and Lori Lipski of Viking Penguin; Jo-Ann Flora, Academic Dean of Stonehill College; Maureen Raymo.

CONTENTS

INTRODUCTION

In 1900 Henry Adams, American historian, grandson and great-grandson of presidents, sixty-two years old, visited the Paris Exposition, a great world's fair celebrating the end of Adams's century and the beginning of a New Age. Again and again he was drawn to the Gallery of Machines, where forty-foot-high dynamos spun at vertiginous speed, scarcely humming as they generated huge quantities of a silent, invisible new force—electricity. Adams did not quite know what to make of the dynamos. He recognized that he was in the presence of something historically important, something that might be the central force of the coming century, but he was baffled by his inability to find in these machines anything recognizably human. He had spent fifty years educating himself. He had published a dozen volumes of history. But in the hall of the dynamos he was reduced to ignorance by forces that he could neither see nor understand. The great wheels spun, power surged through wires, and there was nothing he could detect with his senses. It was a profoundly humbling experience.

Adams turned for enlightenment to his friend Samuel Langley, a respected physicist and aviation pioneer. From Langley's tutorials in the Gallery of Machines Adams came away no more comprehending the great dynamos than before. He could recite the law of magnetic induction, he understood the principle of conservation of energy that linked the dynamos to the hissing steam engines outside the hall, and he could see the visible effects of the electrical currents in the heat, light, and motions they produced. But none of this helped him understand the *meaning* of the machines. "Nothing in education is so astonishing," he wrote, "as the amount of ignorance it accumulates in the form of inert facts."

It was not until Adams had connected the dynamos with his own experience that they became comprehensible. Adams was a historian, concerned with the forces that shape human culture. On his visit to France in the year 1900, two things in particular captured his interest and moved his emotions—the dynamos at the Paris Exposition, and the Gothic architecture of Chartres and Mont St.-Michel. In an essay entitled "The Dynamo and the Virgin" (a chapter in his autobiography, *The Education of Henry Adams*), he explores the motivations that drove an earlier civilization to construct soaring structures dedicated to the Virgin and moved his own time toward the harnessing of mechanical power for material enrichment. Like the cathedrals, *the dynamos embodied human aspirations for order and security in a world of chaos*. The meaning of the dynamos, as perceived and at last understood by Adams, was not electrical but historical, and no amount of inert scientific facts provided by Langley made that history real. Only Adams's personal experience—of the humming machines and luminous cathedral vaults—held the key to understanding.

Many of us stand before the enterprise of science as Henry Adams stood before the dynamos, impressed by obvious achievement but uncomprehending, deferential but skeptical. We know the material prosperity of our culture derives from scientific knowledge, and we know that medical science is the basis for our long and healthful lives. Yet we are apt to feel uneasy that so much of our physical and material lives is hostage to technologists and physicians, even as people in the Middle Ages were beholden to priests for their spiritual well-being. Scientists, like clerics of the medieval church, practice seemingly arcane arts, dealing as if by magic with powerful hidden forces. They tease and conjure welcome benefits from nature on our behalf, but by incantations and instruments we do not understand. No matter how well we are educated, we find ourselves cut off from science by a gulf of incomprehension. This feeling of separateness in the face of scientific knowledge may explain our attraction to such pseudosciences as astrology and parapsychology, which for all their arcane mysteriousness at least seem to connect directly to our personal lives. Further, we are bothered by a vague sense that although science may explain the *what* of the world (a compilation of inert facts), it does not provide the *why*.

It has now been more than thirty years since British scientist and novelist C. P. Snow created a stir among educators with his idea of the "two cultures." According to Snow, "scientific culture" and "literary culture" comprise independent worlds, separated by incomprehension, even by hostility and dislike. Scientists have nothing to say to those who practice or study the humanities—and vice versa. Each culture has its separate vocabulary and agenda (as did Henry Adams and Samuel Langley), and each is impoverished by ignorance of the other. Stung by Snow's challenge, educators tried to patch up the split. Curricula were changed to expose students of the humanities to the sciences, and to make young scientists more sensitive to the arts. Some of these changes were scattered by the educational upheavals of the late 1960s; others continue to limp along ineffectually. More is required than mere tinkering with curricula: The gulf between the two cultures remains as wide as ever.

A few years ago, Alan Bloom, a philosopher of political thought at the University of Chicago, reminded us that the chasm between the cultures has not been bridged. His best-selling *The*

Closing of the American Mind is a querulous and provocative attack on contemporary higher education. He finds little to admire in colleges and universities, and among many failings he includes the mutual incomprehension of the sciences and arts. Bloom concurs with Snow's diagnosis, and takes it a step further: Intellectual life as practiced and taught in our institutions of higher learning has been "decomposed" into not two but three mutually exclusive camps—the natural sciences, the social sciences, and the humanities. Of these three, Bloom sees the natural scientists as aloof and alone, utterly confident of the value of their work, and sublimely indifferent to the rest of the academy (except when political activity is required to ensure the integrity of their departments or financial support for their work). According to Bloom, natural scientists dismiss the social sciences as "imitations" of real science, and consider the humanities—respectfully, of course—as a kind of day-care center where those who ask childlike questions (*Is there a God? Are we free? What is the good society?*) are amused while the adults—the scientists—go about the grown-up work of discovering nature's laws. The natural scientists, Bloom says, are the elitists of the academy, secure in their belief that the only real knowledge is scientific knowledge. It is inconceivable, he asserts, "that a physicist *qua* physicist could learn anything important, or anything at all, from a professor of comparative literature or sociology." And Bloom's gloomy assertion is probably true. The academy is fragmented, as culture is fragmented, and all the king's horses and all the king's men will not put it back together.

If professional academicians are afflicted by mutual incomprehension, then what of the rest of us, forced to live in a scientific culture we do not understand? This is how British biologist Lewis Wolpert describes public attitudes towards science: "Present attitudes towards science seem to indicate both ambivalence and polarization. While there is much interest and admiration for science, there is also a deep-seated fear and hostility. Science is perceived as materialist and dehumanizing, arrogant and dangerous. Its practitioners are a band of cold and unfeeling technicians wielding power without responsibility. Reductionism is suspect and uncomfortable, sabotaging all the mystery and wonder of life. The threats of nuclear war and the genetic manipulation of embryos loom large." Wolpert believes that these attitudes are misrepresentations, caused by a failure of scientists to communicate the true nature of their craft, but I believe they are a fair assessment. The public are not merely misinformed. Science *can* be dehumanizing, arrogant, and dangerous. Its practitioners *are* all too often grim, white-coated technicians wielding power without responsibility. Reductionism *is* suspect and uncomfortable, and *does* sabotage the mystery and wonder of life. And who will deny the overarching threat of nuclear war and genetic engineering?

To remedy this unfortunate state of affairs, Alan Bloom offers nothing more than the doddering suggestion that we all read the Great Books and share (at least) that common ground. It is easy to agree with Bloom that graduates of the university should know Thucydides, Shakespeare, Newton, and Darwin. It is less certain that a shared experience of a pre-twentieth-century corpus of knowledge will humanize science or make it any more comprehensible to the general public. For one thing, science no longer produces Great Books; science has become a very different sort of enterprise from what was practiced by Newton and Darwin or, for that matter, by Samuel Langley. Reading Great Books will not help students of the arts understand what science is today, nor will reading Thucydides or Shakespeare help scientists communicate the meaning of their research. The gulf between the two cultures is immense and it will likely endure. The broken egg will not be put back in its old shell.

Then what can we do to make the incomprehensible comprehensible? How can we discover the soul of science—the understanding that will connect the grand enterprise of science to our personal lives in a satisfying way? The solution, I think, is to follow Henry Adams and look for the soul of science in our own experience. These essays do not attempt to *explain* science, but to *experience* it, to discover the ways in which science enriches our personal lives, not materialistically but spiritually. I draw threads of connection between scientific knowledge and enduring human preoccupations: Personal identity. Sex. Death. Love. Violence. Power. Helplessness. I am convinced that any authentic attempt to understand ourselves must begin with what science has revealed about human nature; I am equally certain that science alone cannot provide ultimate understanding.

Perhaps no ultimate understanding is possible. But one lesson of science seems clear: The age of petty miracles is past. We can no longer look upon ourselves as the favored children of gods. The universe, we now understand, is vast beyond our reckoning, and we are ordinary, perhaps even typical, fragments of that universe. What is required, then, is the courage to accept our cosmic mediocrity and to recognize our oneness with creation, the courage to walk a tightrope between arrogance and despair. We are as special as the universe is special; we are as common as the universe is common. It is as Augustine of Hippo said, "There is but one miracle, and that miracle is creation."

If science is a mirror to creation, then it is also a mirror to ourselves. In these essays I have looked into the mirror of science for some glimmering reflection of the human spirit. If the essays have value to anyone other than their author it is because the experiences I describe are commonplace—not only to scientists, but to anyone who has looked in wonder at the night sky, heard the cry of a loon, felt the subtle hungerings of sex, or worried fretfully about the invisible threat of a virus or radiation. I have sought the *aesthetic, historical,* and *moral* connections between science and human experience that Henry Adams found in the gallery of the dynamos at Paris, when at last he made the connection between the vertiginous machines and the cathedral of the Virgin at Chartres.

At the end of this book, I will return to the theme of the Virgin and the dynamo, but I will choose from a painting of the fifteenth-century Flemish master Robert Campin a rather more modest artifact of technology than the whirring forty-foot-tall dynamos that Adams admired in the Gallery of Machines at Paris. Those machines have lost something of their attractiveness as symbols of security in the midst of chaos. We are less confident now than in Adams's time that science can be an instrument of moral perfection, and we have learned the hard way about the dangers of mega-technology. Scientific knowledge of the world both enriches and humbles. In place of the dynamo, I will choose a humble mousetrap, perhaps the proverbial better mousetrap, to represent our hesitant, tremulous hold upon truth.

THE VIRGIN AND THE MOUSETRAP

CHAPTER I

Tales Told by Starlight

When I was a child I owned a picture book that told the story of Christopher Columbus. Several of its illustrations are still vivid in my memory. One showed Spanish caravels, pennants flying, sailing off the edge of a flat earth into the mouth of a waiting monster. This supposedly illustrated the prevailing view of the earth's shape among Columbus's contemporaries. Another showed Columbus standing before Queen Isabella with an apple in his hand. "I believe the earth is round like this apple," he is saying. "Give me ships and crews to sail them and I will prove it." And so it was that one more child learned the myth that Columbus was the first to imagine that the earth was round.

Columbus did not need to convince anyone that the earth was a sphere. All geographers since the time of the Greeks knew the earth was shaped like a ball, and Isabella and her advisers were certainly well informed on this matter. Nor was Columbus the first to imagine he could reach Asia by sailing westward across the Atlantic. Aristotle reportedly said it could be done, and more than a thousand years before Columbus the Greek geographer Strabo recorded that sailors from the Mediterranean had attempted the crossing. But none of this detracts from Columbus's achievement. To have launched out onto the uncharted western ocean in three tiny ships required exceptional courage, no less than that required by the first astronauts who embarked for the moon. Columbus returned from his voyage convinced he had exploited the sphericity of the earth for the glory of Spain. The astronauts returned from the moon with photographs of a blue and white planet suspended in the black of space, more perfectly round than any apple.

Even with those photographs before us, certain erroneous views persist regarding the shape of the earth. Schoolbooks frequently exaggerate the roughness of our planet's surface. It is difficult to grasp how smooth the planet actually is. We talk of the "mighty" Andes and the "towering" Himalayas as if they constituted considerable bumps on the face of the globe. But the earth's surface is rough only on a human scale. Put a scaled-down Mount Everest on a bowling ball and it would be less than a hundredth of an inch high, about the thickness of a piece of tissue paper. If the surface of a bowling ball were modeled to represent the surface of the Earth exactly, from the highest mountain peaks to the deepest ocean trenches, our eyes would not detect nor our hands feel the roughness, and you could still roll a fair game of strikes. The blue and white planet we see in the photographs from space is dappled by sea and cloud. But even the oceans and the atmosphere are of gauze-like dimensions. On a bowling ball, the atmosphere would be no thicker than a few sheets of paper. Dip the bowling ball into a tub of water, shake it off, and the film of damp clinging to its surface is sufficiently thick to represent the oceans. A cosmic giant who picked up the earth would scarcely be aware that he was holding in his hands anything other than a perfectly smooth, perfectly spherical rock. And, of course, that's exactly what it is. Oceans, atmosphere, mountains, and valleys are considerable only by comparison to ourselves.

To imagine the planet as it really is, we must escape the limitations of human scale; we must cease to live in the world of ordinary perception and enter the realm of imagination. The geographer Eratosthenes of Alexandria in the third century B.C. successfully measured the earth's size by using mathematical reasoning and observations of shadows cast by the sun. But he had first to imagine in his mind's eye a smooth, spherical earth on which even the highest mountains and deepest valleys had shrunk to insignificance. With ink upon papyrus he drew a geometrical circle and said, "This is the earth." In doing so, he transcended common sense and created a new method of discovery. Today we call his method science.

Science is not a collection of facts, nor is science something that happens in the laboratory. Science happens in the head; it is a flight of imagination beyond the constraints of ordinary perception. This was Eratosthenes' achievement. And this, too, was Columbus's achievement. He was not the first to conceive that the earth was round, nor was he first to assert that Asia could be reached by sailing westward. But he successfully freed himself from limitations of the human scale. It is one thing to know something as schoolbook information, as Isabella's advisors presumably knew of the sphericity of the earth; it is something else again to fully comprehend a thing in the mind's eye. Columbus's concept of a spherical earth was sufficiently clear to sustain him on one of the great voyages of discovery. When the horizon stretched endlessly flat before him, he nurtured the image of a round, geometrical earth. When his sailors asserted the imminent danger of falling off the edge of the world, he saw the curve of the sea folding back upon itself. In this he was more than a navigator, more than Admiral of the Ocean Sea. He was a man of imagination and of science.

An old proverb says that a journey of a thousand miles begins with a single step. It is literally true that the journey of Columbus began with a single step, a step taken long before the great navigator set out from Seville. Sometime in the third century B.C., a surveyor in the service of one of the pharaohs walked with measured stride from the city of Alexandria at the mouth of the Nile River and southward along the bank of the river to the village of Syene, counting his paces along the way. His first step was the beginning of the exact sciences. Later, Eratosthenes used the measured distance from Alexandria to Syene (5000 stades in Greek measure, or about 500 miles), together with differences in shadows cast by the sun in the two places, to calculate the size of the earth. Once the diameter of the earth was known, Aristarchus of Samos worked out the distances to the sun and moon from certain observations of the moon at half phase and in eclipse. In 1838, the German astronomer Friedrich Bessel triangulated the distance to nearby stars using the diameter of the earth's orbit (twice the distance to the sun) as his base line. Today we use the distances to nearby stars as yardsticks for determining the distances to more remote celestial objects. And so the journey continues, step by step, taking us ever deeper into the realm of the stars and galaxies. Calculations and measurements have been refined since the time of the Greeks, but the measured stride of a certain unknown employee of the pharaoh was the beginning of our scientific knowledge of the cosmos.

The distance to the most remote galaxies is a billion trillion times greater than the distance from Alexandria to Syene, and a million billion trillion times greater than the length of the surveyor's stride. Not Eratosthenes, nor Columbus, nor Friedrich Bessel could have imagined the extraordinary dimension of the universe that subsequent investigations have revealed. There are as many stars in the Milky Way galaxy as there are grains of salt in 10,000 one-pound boxes of salt; our sun with its family of planets is a typical "grain." With our largest telescopes we can see more galaxies than there are boxes of salt in all of the supermarkets of the world, and among them our Milky Way galaxy appears utterly typical. Within that almost uncountable plurality of worlds we too, with our dreams and fears, are likely to be neither the greatest nor least achievement of creation. Astronomers sometimes speak of "the mediocrity principle." The principle can be stated something like this: The view from here is about the same as the view from anywhere else in the universe. Or to put it another way: Our galaxy, our star, our planet, and even our own life and intelligence are cosmically mediocre.

The mediocrity principle is not easy to accept. It runs counter to the natural human tendency to imagine ourselves special, central, utterly unique. Indeed, it runs counter to what human beings have believed about themselves throughout most of history. In the western world it has been widely held that the universe was created by God specifically as a domicile for the descendents of Adam and Eve, and that humans are the central measure of all things. Even today some scientists accept a related view, called the anthropic principle (from the Greek *anthropos,* "man"). According to this view, life and mind could only have arisen within a universe of *very precisely limited* physical properties. For example, if the so-called fine-structure constant that governs atomic interactions were even slightly different from its known value, then stars would either burn themselves out very rapidly (allowing insufficient time for the evolution of life) or remain forever cold and dark. The fact that we are here to observe the universe means the universe could not have been other than it is. Our existence, then, determines the nature of the universe we live in, selects one universe out of a myriad of imaginable universes. But the mediocrity principle in its strongest form denies even this special role for humankind.

Contemporaries of Columbus were nudged toward the mediocrity principle when they found civilizations in the New World nearly as advanced as their own. Nicholas Copernicus may have been the first to employ the mediocrity principle scientifically when he displaced the earth from the center of the universe and made it just one more planet circling the sun. Galileo was persecuted for denying that the earth was the fixed center of the universe, and his contemporary, Giordano Bruno, was burned at the stake for, among other things, asserting our cosmic mediocrity. The hard fact is this: Every time we thought our place in the universe was special or central we discovered we were wrong. We thought the tribal village was the center of the universe, and we were wrong. We thought Rome or Jerusalem was the center, and that turned out to be wrong. We thought the earth was central, until Copernicus and Galileo proved it was just another planet. Then, with Bessel and his successors, we discovered that the sun is a typical star in a typical corner of a typical galaxy, a grain of salt in 10,000 boxes' worth of salt.

And there is more. Moon rocks, returned to Earth by astronauts, are very like terrestrial rocks. Photographs of the surface of Mars made by Viking landers, and of the surface of Venus by Soviet Venera craft, might as well have been taken in the desert of Nevada. Volcanoes and ice fields on the surface of Jupiter's moons are identical to corresponding features on Earth. Meteorites that fall onto Earth from outer space contain the same organic compounds that are the basis for terrestrial life. Gas clouds in interstellar space are composed of the same atoms and molecules that we find in our own bodies. The most distant galaxies betray in their spectra the presence of familiar elements. All of the evidence of our space probes and our telescopes suggests that the universe everywhere is not much different from the universe here. Indeed, the entire history of science is an argument for the mediocrity principle—not a proof, mind you, but a persuading accumulation of experience. The mediocrity principle is recurring disappointment raised to the status of a truth. And the mediocrity principle says that we are average.

We reel before this hard lesson. We deny the evidence of reason and assert the attentions of ministering gods. We reject astronomy and embrace instead the archaic consolations of astrology, with its implication of an intimate, personal bond linking our lives to the stars. We alienate ourselves from the instruments of knowledge because we are frightened by what we learn, and instead take refuge in a universe drawn on a human scale—towering peaks, uncrossable seas—as the bed-bound child arranges imaginary worlds upon his counterpane.

To say that we are mediocre is not to say that we are insignificant (and here, perhaps, we are at the crux of the alienation of culture from science). Rather than demeaning us, the mediocrity principle establishes our worth as equal to that of the universe. If what is here on earth, in its prodigious variety, is typical of what is elsewhere, then the universe is a rich and ample place. And if we, through imagination, participate in that prodigality, conserve it, treasure it, then we embellish ourselves. By escaping the human scale we become more fully human.

In 1986 I traveled to Australia to look at Comet Halley. The comet was worth the long journey, but even more exciting was my visit to the Australian observatories on Siding Spring Mountain. The setting for this great scientific facility is spectacular. The observatories are ringed by jagged volcanic peaks of the Warrumbungle Mountains, mantled with gum trees and susurrous with the songs of bush birds. In the grassy valleys at the mountains' base kangaroos and emus graze at dawn and sunset. Beyond the mountains the iron-red outback reaches to the far horizon.

In the largest of the observatory domes is housed the Anglo-Australian 3.9-meter telescope, one of the largest telescopes in the southern hemisphere. Another building contains the Australian Advanced Technology Telescope, a pioneer in the new economies of cost that can be achieved in telescope design by exploiting the capabilities of high-speed computers. Other domes hold smaller instruments. It was in the dome that houses the British 1.2-meter Schmidt telescopic camera that I found what I was looking for. The Schmidt telescope was engaged in a systematic photographic survey of the southern sky. The photographs are made on glass plates coated with emulsions created especially for astronomy. The glass plates are not much thicker than a thumbnail; they are thin so they can be bent to match the curved focal surface of the telescope. Each plate is about the size of a newspaper page and covers a part of the sky equal to the size of my palm held at arm's length; it would require almost eighteen hundred of these photographs to cover the entire sky. A typical exposure lasts about an hour, during which time the telescope must be moved with extreme accuracy to compensate for the turning of the Earth. The telescope records fainter stars and more distant galaxies than have ever been seen before. Each photograph contains between one million and ten million visible images—sharp spots and fuzzy spots. The sharp spots are stars in our own Milky Way Galaxy. The fuzzy spots are mostly other galaxies, other island universes that contain as many as a trillion stars apiece. About half of the images on any plate are galaxies. I had the opportunity to examine several contact negatives with a magnifier. In the magnifier, the brightest of the fuzzy spots become spiral galaxies of dazzling detail. Many of the fuzzy spots are interacting galaxies—two or more great star systems locked in a spiral-distorting gravitational dance. On each plate there are recorded as many as a million galaxies. Each galaxy contains hundreds of billions of stars. Many of those stars, like our sun, have planets. With the magnifier, I examined in a few minutes more worlds than my mind was capable of imagining.

What does it mean, this extravagant profusion of worlds? I had traveled halfway around a planet to visit Siding Spring Mountain, following westward the old trajectory of Columbus. It seemed an enormous distance. Traveling at the same speed—the cruising velocity of a 747 jetliner—it would take two trillion years to reach the nearest of the galaxies recorded on the photographs, and to reach the most distant of the galaxies would require a time greater than the age of the universe. Often when I am teaching the astronomy of galaxies one of my students will say to me, "It makes me feel so small." I disagree. To be mediocre is not to be small.

Every galaxy whose image is fixed on a photographic plate has become a permanent part of the human imagination. Through the agency of the telescopes on Siding Spring Mountain, and others like them throughout the world, the human imagination has gone out to embrace the distant galaxies. We are physically small compared to the galaxies, but our minds encompass the universe.

In cosmic time, the white domes on Siding Spring Mountain sprang up as quickly as the mushrooms of my beloved New England woodlands. And they will disappear as quickly. Like mushrooms, they are a part of the story of life in the cosmos, a cosmos that here—and probably elsewhere—has achieved self-consciousness. As I examined the photographic negatives on Siding Spring Mountain, this thought occurred to me: We are almost certainly not the brightest thing the universe has yet thrown forth, but we are certainly not small. Our imaginations are billions of light years wide. We can justifiably say with Shakespeare's Miranda, "O, wonder! How many goodly creatures are there here! How beauteous mankind is! O brave new world that has such people in't!"

CHAPTER 2

Van Gogh's Night

"If the stars should appear one night in a thousand years," wrote Ralph Waldo Emerson, "how men would believe and adore, and preserve for many generations the remembrance of the city of God which had been shown. But every night come out these envoys of beauty, and light the universe with their admonishing smile." Emerson was a New Englander and should have known better than to say that the stars come out "every night." The average cloud cover in New England year-round is about 60 percent. Even in the least cloudy month, October, the Transcendental philosopher of Concord had only about a fifty-fifty chance of seeing stars, although that was frequent enough, one must suppose, for the heavens to disguise their transcendence in the cheap cloth of familiarity. Things could be worse; in Ireland, where I spend my summers, we are grateful for one starry night in ten. Or better; in certain parts of Arizona the sky is clear about 80 percent of the time, and on top of Mauna Kea volcano in Hawaii it's about the same. Arizona and Mauna Kea are homes for some of the world's biggest and best observatories. The skies in those places are almost preternaturally clear. The average cloud cover for the entire globe is about the same as it is for New England: 50 or 60 percent. But what if cloud cover were 100 percent? What if the stars failed to reveal themselves even for Emerson's one night in a thousand years? How would the intellectual history of the human species have been different on a cloud-wrapped planet?

The question may not be altogether frivolous. Clouds consist of tiny drops of water, so fine they float in billows. When rising air cools, water vapor in the air condenses into liquid drops. Each drop forms upon a solid nucleus, a particle of sea salt, wind-borne soil, volcanic dust, or smoke from forest fires. All of these things—rising air, water vapor, condensation nuclei—depend upon climate; that is, they depend upon the overall balance of solar energy and its distribution by winds and ocean currents. The system of climate is immensely complicated and there are many forms of feedback. It may be that Earth's average amount of cloud cover is more or less self-regulating. But it is also possible that a modest change in any of several variables might have yielded a more Venusian planet, totally wreathed in vapor. On such a planet no one would observe the starry night, the changing face of the moon, the dance of the planets, the geometrical exquisiteness of a solar or lunar eclipse, or the fierce unblinking eye of the sun. These are the very things that excited the scientific instinct in the minds of our ancestors.

The stars are not just Emerson's "envoys of beauty"; they are also envoys of order. No other part of the natural environment so clearly betrays regular periodic phenomena as the heavens. Anthropologist Alexander Marshack has argued that certain man-made scratches on Ice Age artifacts record the changing phases of the moon, and that these are the earliest examples of symbolic notation, a kind of proto-number system. Historians Giorgio de Santillana and Hertha von Dechend contend (in a book called *Hamlet's Mill*) that all of the great myths of the world have their origin in the regular, periodic behavior of celestial bodies, most especially the recurring seasonal migrations of the sun. Many other commentators have stressed the connection between observations of the heavens and the early development of scientific thought. As Jacob Bronowski pointed out, the stars might seem to be improbable objects to have aroused such curiosity. The human body is closer at hand and a more obvious candidate for systematic investigation. But astronomy advanced as a science before medicine; indeed, early medicine turned to the stars for signs and omens. The reason is clear: The regular motions of celestial objects lend themselves to mathematical description. Behind the apparent chaos of terrestrial experience, vexing and capricious, the stars proclaim the rule of law. The city of God, revealed in the starry night, is stark, reassuring, ordered.

To my mind, the greatest scientific work of antiquity was Aristarchus' measurement of the sizes and distances of the sun and moon, described in a book that can hold its own with the best scientific treatises of today. And with what stunning conclusions! The sun is vastly bigger than the earth; the distances to the sun and moon, and therefore to the stars, are almost unimaginably large; the earth is an insignificant mote of dust in the gaping immensity of the universe. It's all there, for the first time, in Aristarchus' closely reasoned pages, every aspect of scientific method: abstraction, calculation, and quantitative measurement, to which is added flashes of insight excited by the changing phases of the moon and the coppery adumbration of the moon in eclipse. Here are the first intimations of our cosmic mediocrity. Together with his contemporaries or near-contemporaries, the geographers, astronomers, and mathematicians of the Alexandrian school, Aristarchus invented science. He distilled science from the stars. Science was the child of the uncloudy sky.

On a cloud-shrouded earth the rise of the human species to scientific civilization would almost certainly have been delayed. Delayed, but not forestalled forever. The survival value of science (and its attendant, technology) is such that sooner or later the inhabitants of a White Planet would have developed the vehicles to lift themselves above the clouds, rising perhaps on columns of flame as do our astronauts. We can imagine their first view of the universe beyond the clouds—the beckoning stars, the Milky Way, the incandescent orb of the sun, the pale winking moon, planets, comets, solar and lunar eclipses—rhythms unhidden, the music of the spheres revealed, the mathematical rule of law, so laboriously learned in the terrestrial environment, in the heavens made crystal clear.

For those of us who live within twenty-five miles of a city, summer vacation may be our only chance to rediscover the starry night. No matter how many times it has happened before, there is still a shock of surprise on that first clear night of the season, on a lake shore in Maine or a campground high on a western plateau, when we lean back our heads and exclaim with the poet Gerard Manley Hopkins:

> *Look at the stars! look, look up at the skies!*
> *O look at all the fire-folk sitting in the air!*
> *The bright boroughs, the quivering citadels there!*
> *The dim woods quick with diamond wells;*
> *the elf eyes!*

It is a pleasant coincidence that summer vacation coincides with the most richly star-drenched skies of the year. In the evening hours of August and early September we look out from earth toward the glittering center of our galaxy, an overbrimming diamond well of stars that we see as a drapery of milky light hung across the sky. As I write, I am thinking particularly of an evening not long ago when I was far from city lights under a sky of crystalline clarity. The earth was tented with stars, stars so numerous they appeared as a continuous fabric of light. The Milky Way flowed like a luminous river from north to south, banked with dark shoals, eddied in glittering pools. Our sister galaxy in the constellation Andromeda was visible to the naked eye, a blur of light from a trillion faraway stars. Meteors flashed like fireflies. Such skies never fail to excite the imagination. Certain constellations—Orion or Ursa Major—are perhaps the oldest surviving inventions of the human mind. The depth and beauty of the night inspired religious and philosophical speculation. Science and mathematics had their origin in the questions posed by the night, by the lopsided circlings of sun and stars and the movements of the planets on their shuttlecock courses. And now, for a great portion of humankind, the starry night is gone, obliterated by artificial light and haze. It is a measure of the degree to which we have polluted our skies that while almost everyone has heard of the Milky Way, surprisingly few people in the developed countries have seen it. These are the first cultures anywhere on earth, at any time in history, for whom the Milky Way is not a prominent and inspiring part of the natural environment. Other more subtle lights in the night sky—the Great Galaxy in Andromeda, the double cluster of stars in Perseus, and the "Beehive" star cluster in Cancer are examples—once easy naked-eye sights, are now available only to those with access to the darkest, clearest skies.

Our present system for defining the brightness of stars goes back to the astronomer Hipparchus, who lived in the city of Alexandria at the mouth of the Nile River in the second century B.C. The dozen or so brightest stars in the sky he called stars of the first magnitude. The faintest stars he could see (in those days before telescopes or any kind of optical aid) he called stars of the sixth magnitude, a category that included thousands of tiny pinpoints of light at the limit of vision. Other stars he assigned to the appropriate intervals between. No sixth magnitude stars are visible from the city of Alexandria today, nor indeed from any city. On an extraordinarily clear night a city stargazer might see stars of the second or third magnitude, a few hundred dots of light sprinkled inconspicuously across the sky. A more typical urban night might reveal stars of the first magnitude only. Year by year, more and more of the starry night is subsumed in the glow of streetlights or obscured by atmospheric pollution. But away from city lights, in the remote fastness of a summer retreat, it is still possible to recover the sky as the astronomer Hipparchus or the poet Gerard Manley Hopkins saw it, awash with stars of the sixth magnitude, glittering fire-folk displayed from horizon to horizon in teeming quivering citadels.

What can be done to halt the erosion of the night? Probably not much. Increasingly, the business of civilization proceeds around the clock and nighttime activity requires artificial light. The requirements of security in a dangerous and uncertain world bring artificial light into the suburbs and countryside. The sickly orange glow of mercury-vapor lamps displaces the Milky Way, the subtle zodiacal light, and the shimmering aurora. Amateur and professional astronomers struggle to make governments and private citizens aware of the aesthetic, spiritual, and scientific importance of dark skies. Modest victories have been won on behalf of darkness, but one cannot be optimistic about long-term prospects. The day is perhaps not far off when stars will be something city-dwellers read about in a book, as even now, for many people, the Milky Way is something they have heard of but not seen. As Hopkins's "bright boroughs" become increasingly remote, not even a summer's evening on a Caribbean isle or the high dark rim of the Grand Canyon will proffer the undimmed starry night. Technology, the child of darkness, has become its slayer.

No artist so vividly captured the fulgent magic of the undimmed starry night as the nineteenth-century Dutch painter Vincent van Gogh. Two of his paintings especially, *Starry Night on the Rhône* and *Starry Night*, give credence to the inspiring force of Emerson's one starry night in a thousand years and to Hopkins's "bright boroughs." These vertiginous layerings of pigment portray a sky uncorrupted by the vaporous exhalations of civilization, igniting fierce fires of imagination.

Van Gogh's tumultuous nighttime paintings have sometimes been thought to be products of the artist's madness. In fact, they are breathtakingly sane. Consider but one of these works, *Road with Cypress and Star*. Three celestial objects are represented: a crescent moon to the right of a flamelike cypress tree, a bright star, and another less-bright star near the horizon on the left. Is this skyscape a product of the artist's imagination, or was it inspired by an actual configuration of heavenly bodies? Astronomers Donald Olson and Russell Doescher argue for the latter interpretation. *Road with Cypress and Star* was painted at Saint-Rémy, in southern France, near the end of van Gogh's year-long stay at an asylum in that town. The artist left Saint-Rémy on May 16, 1890, for Auvers, near Paris, where he committed suicide two months later. Using a computer to reconstruct celestial positions, Olson and Doescher searched backwards from the date of van Gogh's departure from Saint-Rémy for a likely arrangement of stars and moon. A crescent moon occurred on April 19. The brilliant "evening star" Venus was near the moon on that date and little Mercury was near the horizon. The configuration of the three objects in the sky, as revealed by the computer, was strikingly similar to the objects in the painting, except that the order of the objects is reversed left to right and the moon's crescent is tipped in a way that never happens in the real sky. The two astronomers are convinced that the April 19, 1890, conjunction of Mercury, Venus, and the moon was the inspiration for van Gogh's painting.

Astronomer Charles Whitney and art historian Albert Boime have analyzed van Gogh's *Starry Night on the Rhône* and *Starry Night* from the astronomical point of view. They used a planetarium to reconstruct past skies, and traveled to France to observe the sky from the places where van Gogh experienced it. They found elements of scientific realism in both paintings. In *Starry Night on the Rhône* the Big Dipper is easily recognized, although the artist has placed the Dipper, a northern constellation, in the southwest. Boime purports to find Venus and the constellation

Aries among the stars of *Starry Night*. In the spiraling swirls of *Starry Night* both Whitney and Boime see the influence of Lord Rosse's 1845 drawing of the spiral galaxy M51, known as the Whirlpool Galaxy. They guess that van Gogh may have seen a representation of Lord Rosse's sketch in the works of the French astronomical popularizer Camille Flammarion. As Albert Boime makes clear, van Gogh was keenly interested in astronomy, cartography, and science in general. He was an exact observer of the night. In a letter to his sister, the artist says that "certain stars are citron-yellow, others have a pink glow, or a green, blue and forget-me-not brilliance." Stars do indeed exhibit these colors, but only to a perceptive observer. On this and other evidence, Whitney concludes that van Gogh had excellent night vision.

But for all the care and perspicuity of his observations, van Gogh's skies are unlike any I have seen. His stars are whirling vortices of color, not cold points of distant light. Blue-black night yields in the paintings to torrents of yellow and green. Moons burn with the vital intensity of suns. Space seethes with the energy of flame. Madness? Insanity? Surely, one might guess, these gaudy celestial billowings, these aureate constellations, are inventions of the artist's dementia. But art historian Ronald Pickvane thinks not: "Between his breakdowns at the asylum [van Gogh] had long periods of absolute lucidity, when he was completely master of himself and his art. That his mind was informed and imaginative, interpretive and highly analytical can be seen in the way he assessed his own work." By finding elements of astronomical realism in van Gogh's paintings, Olson, Doescher, Whitney, and Boime confirm that these provocative visions of the night are not the products of madness. But neither are they literal representations of the sky. They represent, in Pickvane's words, "an exalted experience of reality."

From the barred window of his room at the asylum van Gogh had unobstructed views of the night sky. His insomnia gave him ample opportunity to observe the stars. What he put onto canvas was more than what he saw, and more than what a computer or planetarium can reconstruct. In one of his letters he wrote: "I should be desperate if my figures were correct . . . my great longing is to make these incorrectnesses, these deviations, remodellings, changes of reality that they may become, yes, untruth if you like—but more true than literal truth." Van Gogh the artist did nothing more than what the scientific astronomer does every day: he looked beyond the mere data of experience—beyond the cold specks of light arrayed as glittering diamonds on the jeweler's black cloth—into giddy depths of a haunting, diluting, and even terrifying dimension.

We look into the unclouded starry night and see perhaps more than we want to see. The painter Georges Braque said: "Art is meant to disturb. Science reassures." The color-splashed, starry vortices of van Gogh's nighttime paintings certainly disturb. They disturb because they evoke something that in our less exalted way we also recognize as truer than literal truth. The whirlwind stars of *Starry Night* draw us up, out of ourselves, into a beautiful, uncertain universe, a universe in which the individual, chastened by recognition of his cosmic mediocrity, struggles to find security and meaning. Knowing that these wildly turbulent images contain elements of mathematical order is only mildly reassuring.

CHAPTER 3

Dreamtime

Comet Halley. Heralded by hype. By a fanfare of overblown media rhetoric. By two-inch headlines. By hoopla. By hawkers of shabby telescopes ("comet-catchers"). By breathless television meteorologists (who should have known better). People waited for a great fire-tailed beast to sweep across the sky. Waited for whoosh and glitter. And saw nothing.

The 1985–86 apparition of Comet Halley was for many people the astronomical debacle of the century. It left a trail of wounded enthusiasm and broken hearts. Disappointed would-be comet-gazers turned back to their television sets and decided—this time for good!—to leave the sky alone. But the story of the comet is worth telling once again, precisely because it was such a subtle thing. The gifts of the sky are not extravagantly given, especially in an age of urban lights and industrial pollution; they must be earned. Like van Gogh's stars, we must endow the comet with our own aureole of glamor. We must hear the whoosh with an inner ear. In the course of my life I have seen a thousand bits and pieces of Comet Halley. The Eta Aquarid meteor shower of May and the Orionid meteor shower of October are probably caused by debris (sand-sized bits of rock colliding with the earth's atmosphere, burning up in streaks of luminous vapor) shed by Comet Halley in previous circumnavigations of its orbit. During forty years of observing "shooting stars," I have visually collected a few ounces of the comet's mass. But there is a trillion tons of rock and ice still in the comet. It flies on a long elliptical orbit that brings it close to the sun once every seventy-six years. Most of us are given only one chance to see the comet. The visitation of 1985–86 was my chance.

My search for the comet began in September 1985, when according to predictions it would first come within reach of my fourteen-inch telescope. At that time, Comet Halley was crossing the gulf of space between the orbits of Jupiter and Mars, inbound on its once-in-a-lifetime visit to the inner solar system. It was an object of the twelfth magnitude, which means it was twenty million times less bright than the faintest star that can be seen with the naked eye. I knew the comet would be highest in the sky just before dawn. I waited until mid-month, to allow the moon to pass out of the morning sky, and I watched the weather. The night of September 14 promised to be dark and clear. I set my alarm for 4:00 A.M. I think I saw it. I knew exactly where to look, and when I glanced there with averted vision (to exploit the most sensitive part of the eye's retina) I thought I could detect a slight modulation of the darkness. I think I saw Comet Halley, but I won't swear to it. What I saw may have been no more than a wish for the comet. But I did see Comet Giacobini-Zinner that morning, a first-time visitor to the inner solar system, which just happened to be in the same part of the sky, only a degree or two from Halley and at that time brighter than its more famous periodic cousin.

Mid-October brought another moonless "window" to look for the comet. Halley was still farther out in the solar system than the orbit of Mars, but picking up speed in its fall toward the sun. And brightening. The predicted brightness placed it at the tenth magnitude, easily within reach of my telescope. This time my chance of seeing Halley was good. Again I rose before dawn and made my way to the observatory. A cold front had passed during the night and swept the sky clear of every trace of haze. The atmosphere was unusually steady. I swung my telescope to the star Betelgeuse at the shoulder of Orion, then tracked up the giant's arm and out along the club. I knew the comet was very near to the star Chi-2 Orionis, at the tip of the club. That region of Orion is in the thickest stream of the winter Milky Way. The field of my telescope sparkled with twinkling lights, Hopkins's dim woods quick with diamond wells. The object I was looking for was one hundred forty million miles from Earth, moving on a long elliptical track that takes it nearer to the sun than the planet Venus and farther out than Neptune. The nucleus of the comet is a ball of dust and ice, a few miles in diameter. What I hoped to see was not the nucleus but the much larger coma, a cloud of dust and gas that has been released from the nucleus and surrounds it like a halo, glowing in reflected sunlight. I nudged the telescope north and east of Chi-2 Orionis and quickly found the comet. What I saw was a circular blur of light, brighter at the center, fuzzy at the edges, like a smudged star. The nascent tail, if there was one, was hidden behind the coma. The comet drifted like a ghost among a glittering of stars.

The first two weeks of November brought another moonless opportunity to look for Halley. The comet had moved into the constellation Taurus and was steadily brightening. Now it was an object for backyard binoculars, swelling in sunlight and delicately beautiful. But you would not say Halley was conspicuous. You would not have noticed the comet if you did not know where to look and what to look for. It was like the residue left by a water droplet on a mirror, or the dimple made by the foot of an insect on the surface of a pond. Night after night I followed it toward its rendezvous with the Pleiades. In December and January I watched it fall through Pisces and Aquarius toward the sun. There was an intermission as the comet passed perihelion (its nearest approach to the sun), on the opposite side of the sun from Earth, hidden in the sun's light. Then in March, from a dark hilltop near my home in Massachusetts, I saw Halley emerge from the rosy light of dawn and burn like a slow fuse toward Sagittarius, still—in the light-polluted skies of the Boston suburbs—an object for binoculars only.

If the planet Earth could have halted in its orbit in October, Comet Halley would have passed very near. At that time the comet was above the plane of Earth's orbit (the northern side) and North Americans would have had a splendid view, as good as any in the past and worthy of all the media hyperbole. But Earth did not stop; it went on its way, in the opposite direction around the sun, leaving Halley behind. In the spring, after perihelion and my March glimpse from the hilltop, the comet and the planet approached again. But now Halley was below the plane of the Earth's orbit, out of sight for observers in the northern hemisphere.

My first naked-eye view of the comet was through the window of a Qantas 747 airliner 28,000 feet above the Pacific. I was a member of a Comet Halley Tour sponsored by *Sky & Telescope* magazine, heading for Ayers Rock in central Australia with a group of remarkably like-minded people, all keen observers of the comet and knowledgeable about the sky. Most of our group had been observing Halley for as long as I had, or longer (one of us, Clarence Custer, a pioneer amateur astrophotographer, had witnessed the 1910 visitation of the comet). There would be no disappointment among my companions, no wringing of hands or expressions of dismay when the comet failed to live up to its advance billing. These people were realists. They knew that because of the orbital configurations of Earth and comet this was destined to be one of the least spectacular visitations of Comet Halley of all time. They knew what to expect long before they stepped out into dark Australian skies. I was traveling with connoisseurs of blurs, aficionados of night's subtlest lights. All of us were prepared to accept whatever delicate revelations Halley offered.

It was appropriate that Ayers Rock was the primary viewing site selected for the tour, quite apart from the fact that the sky at that place promised to be exceptionally dark and clear. Ayers Rock, like Comet Halley, is one of those natural phenomena that holds a powerful, almost primordial attraction for the human imagination. The geological term for Ayers Rock is an inselberg: an isolated block of stone rising abruptly from a plain. Most inselbergs of the world are resistant granite masses that have been exposed by the erosion of softer, overlying sedimentary formations. Ayers Rock is unique in that it is itself sedimentary. How it has resisted the wasting that flattened the surrounding plains is something of a mystery. It appears to have been set down onto the Australian outback like an overturned bowl. It shimmers in tropic light with vibrant color. Its inexplicable ruddy presence is huge and edifying. Comet Halley has something of that same uniqueness, that same out-of-nowhere magic. The Pitjantjatjara aborigines who live near Ayers Rock call comets Wurluru. Wurluru is a very large man who lives alone and sometimes hurls spears across the heavens. He is ferocious and powerful, greatly to be feared, but not without merciful qualities. The aborigines say that Wurluru should not be looked at for long periods of time or he will cause the eyes to spin around. On the desert near Ayers Rock we refused to heed the aboriginal advice. Even on our first night at the Rock, most of the group were at the observing site until the early hours of the morning. By the time we drifted off to bed all eyes were spinning.

An astonishing variety of optical instruments was directed at the comet: cameras, binoculars, and telescopes of all sorts. The click of shutters, the beep of timers, the whir of drive mechanisms (and the ticktock of one curious cuckoo clock-powered camera platform) resembled the night sounds of exotic desert fauna. The comet rose in air so dark and clear we spotted the coma even before half of its diameter had cleared the distant horizon. Even against the backdrop of the southern Milky Way (the richest part of all the sky) the comet was conspicuous. The tail, however, was poorly developed and difficult to trace for more than a few degrees across the sky. For several nights the comet made its way along the curved body of the Scorpion and through the brightest stream of the Milky Way. Then, in our second week in the southern hemisphere, it broke free of the Milky Way into the pitch black sky of the constellation Lupus. To my eye, the comet appeared brightest on the night of April 10, when we were in Tahiti on our way home, but such perceptions can be subjective.

There were plenty of objects in the sky by which to gauge the comet's brightness, many of them new to those of us who were seeing the southern sky for the first time. Star clusters. Nebulas unfolding like colored paper flowers. The Large and Small Magellanic Clouds, companion galaxies to the Milky Way. Among these glittering prizes Halley moved, relucent, majestic, history-laden. During our time in Australia it traversed the sky more rapidly than at any other time in its apparition. Pictures taken only hours apart made a fine stereoscopic pair, giving the illusion of a third dimension (depth) against the background of distant stars. The comet seemed a tranquil object, a ghostly, crewless galleon adrift on a misty sea, but that too was an illusion. Three weeks earlier the comet had been visited by an armada of spacecraft. The Soviet *Vegas* 1 and 2 and the European *Giotto* craft had plunged within the coma and revealed the nucleus to be an active sputtering mass of volatile materials that respond violently to heating by the sun. Jets of dust were expelled from the nucleus with explosive intensity. And the spacecraft revealed another irony: The luminous object that we so patiently courted from the murmuring night-desert of Australia was the halo of an object that was actually coal black. The nucleus of Comet Halley is encrusted with ebony dust.

There was a special magic about observing Halley from the neighborhood of Ayers Rock. The culture of the Pitjantjatjara aborigines centers upon a body of myth and ritual that had its origin in the Tjukurapa—or Dreamtime—before which the earth did not exist. The aborigines believe that animal ancestors from the Dreamtime created the world by crossing and recrossing a flat featureless desert. Where these creatures moved or rested, forests, rivers, water holes, and mountains came into existence. Many tracks crossed at Ayers Rock (called Uluru by the aborigines), and the paths coalesced to form the stupendous monolith of red stone. The stories of the Pitjantjatjara refer especially to the snake people, the carpet snakes and the poisonous snakes, whose battles and adventures account for many features of the Rock. In the view of the aborigines, the Rock is a concretion of Dreamtime elements. It is sacred because of the way it participates in the origin of the world. Something of Ayers Rock's sacred character is apparent even to non-aboriginals, or how else account for the aura of mystery that many of us felt as we watched the comet from within the shadow of the monolith?

Throughout history and in many cultures, snakes have been a common image for comets. Like Uluru, Comet Halley comes snaking to us from a cosmic Dreamtime, from a time before the earth was made. Scientists are keenly interested in comets because they are believed to preserve unaltered primordial ingredients of the early solar system. The spacecraft that penetrated Halley's coma and tail encountered particles of dust that are as close as anything we have yet observed to the materials out of which the earth was made. Organic chemical compounds detected in the spectra of the comet hint at the possibility that comets bear clues to the origin of life—that we ourselves might be made from comet dust. Halley comes to us from the era of the creation, crossing and recrossing the featureless desert of space, a concretion of primordial elements, a totem of our beginnings.

We watched the comet rise over Ayers Rock, subtle, delicate, but equal through binoculars to the brightest of the globular clusters, spheres of a hundred thousand suns, or to nebulas such as the Tarantula, vast gassy clouds where suns and earths are born. Our fascination with the comet had nothing to do with its apparent brightness or the length of its tail; it had to do with the significance of the comet as an object of scientific study and as a symbol of our long preoccupation with the sky. No one in our group was disappointed that the comet did not present a grander aspect. I have always felt that people who care deeply about the night sky are invariably concerned with origins and history. The names of the stars, the patterns of the constellations, the star lore of the Australian aborigines, and modern astronomy are all products of human curiosity about who we are and where we came from. Even in its subtle, almost tailless guise, Comet Halley was a totem of that quest.

Astronomy is a science of faint lights. The excitement of astronomy lies in the way grand knowledge is distilled from barely luminous blurs in the night sky. One blur is a nebula where stars are born from celestial steamers of dust and gas. Another is the debris of a stellar explosion, a supernova, fat with heavy elements for future planets. Other blurs are star clusters, quasars, galaxies racing outward from the impulse of the Big Bang, or comets bearing hints of the origin of life. These faint lights, imaginatively interpreted, convey to us the secrets of the Dreamtime and confirm our unity with the cosmos. One night in late April, back in Massachusetts, I showed the rapidly retreating comet to a teenaged friend. It was again a telescopic object, receding from the earth, climbing back to the top of its long roller-coaster track. Again it looked like a smudged star and a dim one at that. "Is that it?" my friend asked incredulously. "Yep," I answered. "That's it." He took another disappointed look. "That's useless," he said in the lingo of his generation. And I thought of something Samuel Johnson wrote about poets, something that applies equally well to professional and amateur astronomers: "To a poet, nothing can be useless. Whatever is beautiful, and whatever is dreadful, must be familiar to his imagination: he must be conversant with all that is awfully vast and elegantly little." Comet Halley, Dreamtime messenger, ebony-hearted giver of life, snaky totem of our beginnings, was awfully vast and elegantly little.

CHAPTER 4

The Oldest Question

We are made of star stuff. Astronomers have detected in interstellar nebulas (those vast gassy clouds that drift in space between the stars, from which stars emerge and into which dying stars disperse their substance) the spectral signatures of organic molecules: water, ammonia, formaldehyde, methyl and ethyl alcohol, acetylene, methanimine, and dozens more. Meteorites that fall upon the earth from space bear amino acids, the chemical units that are the backbone of the proteins. And comets—including, especially, Halley, the most intensely studied of all comets—bear within their volatile burden many of the chemical building blocks of life. We are the effervescence of creation, spume cast from the sea of matter onto a welcoming shore: carbon, hydrogen, oxygen, and nitrogen, tossed together into chains and streamers, lashed into animation.

Chemically we are no different from the rest of creation. A few pennies worth of atoms. Our value is not that of gold or platinum, made special by rarity; our elements are among the most common in the universe. Pull us apart into our constituent parts and you would have unexceptional substances: a few vials of colorless gases, a thimbleful of carbon, smidgens of sulfur, phosphorus, and iron. But this conglomeration of ordinary matter is anything but commonplace. Each of us is intensely conscious of being unique. Out of utterly unexceptional substance self somehow arises. There is no greater mystery than this: the genesis of particularity.

"Who am I?" It is the oldest question in philosophy. Socrates asked it. Descartes asked it. Philosophers ask it today. And science may be on the verge of breakthroughs that will change forever the way we understand the question. Two parts of the self are currently open to scientific scrutiny. If I consider myself, the first part was there at the self's beginning, genetically encoded in the single cell (or union of two cells) that would become me: *Homo sapiens*, male, white skin, brown eyes, black hair, a tendency to baldness in middle age. The second part is the treasury of experiences stored in memory: songs heard, visions seen, tastes, smells, pleasures and traumas, available for recall and association. These things together make up the creature that is uniquely "I."

Biologist Francis Crick, co-discoverer of the structure of DNA, once estimated that if the biochemical instructions for building a human being could be placed into 500-page recipe books, more than one thousand volumes would be required. A complete copy of that huge quantity of information is contained in every human cell as DNA molecules. DNA has the form of a spiral staircase. The side rails of the staircase are repetitive chains of sugars and phosphates. The treads are pairs of chemical units called nucleotides. There are four kinds of treads (designated A, T, G, and C by biologists). On the stairway of human DNA there are three billion steps. It is the sequence of treads on the stairway that is the code of life.

Molecular biologists have invented techniques for determining the sequence of nucleotides, and have now embarked upon a remarkable program—the "genome project"—to provide a complete transcription of the human DNA (a thousand 500-page volumes of four-letter code). It is a staggering task, involving thousands of person-years and billions of dollars in resources. Before the year 2000 the entire sequence of three billion nucleotides will be stored in a computer bank. One segment of the sequence is the gene for sex, another for the color of the eyes, yet another for a particular inheritable disease. The code will be available to researchers in medicine, genetics, and evolutionary biology. It will be a complete chemical blueprint for a human being.

In his autobiographical book, *The Double Helix,* James Watson (Francis Crick's collaborator in the discovery of the structure of DNA and now director of the genome project in the United States) tells how he came to regard the helix as the fundamental shape of the DNA molecule. "The idea [of the helix] was so simple," he said, "that it had to be right." A few years ago, the journal *Scientific American* devoted a special issue to the molecules of life, illustrated with stunning computer-generated images of DNA, proteins, hormones, and the other molecules that make us go. These molecules are complex—they are made up of thousands or billions of atoms, represented on the computer screen as points of light—yet their structures are so simple and beautiful that when we look at them we know they are "right." Chemists often build models of molecules with sticks and plastic spheres, but the computer images are more striking than any mechanical model I have seen. On the computer's screen the molecules can be turned, twisted, mated, and modified. One can watch, for example, as a disease-causing antigen binds itself to an antibody, a central event in the body's recognition of the antigen as a foreign organism. And the computer can represent many more atoms than can be reasonably incorporated into a model made with balls and sticks. The electronic molecules glow luminously against black backgrounds, atoms of each element shimmering in its coded color, dim woods quick with diamond wells, the magic of Hopkins's stars writ small.

If I focus on the "atomic" dots of the computer images, I am bewildered by an astonishing complexity of detail, and I wonder that life exists at all. But when I focus on the overall patterns—the geometrical forms and symmetries of the molecules—I am dazzled by what seems to be an almost inevitable simplicity. Every molecule seems miraculously contrived for its task. As I studied the computer images, I had the feeling I had seen them before. And then I knew what it is that I was remembering: the gorgeous stained-glass windows and soaring architectural components of the Gothic cathedrals. An image of the B DNA double helix, shown in cross-section, bears an astonishing likeness to the magnificent rose window at Chartres, and in the webbed vaulting of clathrin protein and the flying buttresses of the sugar-phosphate side chains of DNA I had that same sense of déjà vu. No medieval architect could have raised more fitting structures.

Medieval builders sought to reflect in the visible structures of their cathedrals invisible realities of the world of spirit. The cathedral was an earthly image of the kingdom of God. When we enter a Gothic cathedral we have the sense that every visible component of the structure has a job to do; the Gothic architects achieved a unity of form and function that has seldom been surpassed. Something similar is afoot in the computer images of the molecules of life. Here, too, there is an attempt to represent unseen realities with visible images. Here, too, is an almost mystical vision of a hidden harmony established throughout the cosmos, the unsuppressible capacity of substance to generate self.

Gothic style was determined by medieval materials—stone and mortar (which bear compressional loads only)—and by the twin theological objectives of height and light. From these materials and objectives, Gothic form emerged: webbed vaults, flying buttresses, rose windows. Within the constraints of that necessary style, each cathedral is unique. Chartres, Rheims, Salisbury,

York: Gothic, yet undeniably individual. Within the spiraling helix of the human DNA we discover this same capacity for variation within unity. One of the *Scientific American* images shows the *cro* repressor molecule affixing itself to the DNA helix of a bacterial virus; the *cro* repressor acts to prevent the expression of a gene, and thereby exerts a tiny but significant twist to the thread of life, modifying a minute element of self. On the computer screen the two molecules come together and mate like lock and key. No atom appears in excess. Nothing is wanting. It is said that medieval theologians debated about how many angels could dance on the head of a pin; the glissades and pirouettes of the antigen and the *cro* repressor, and the whirling tarantella of the DNA double helix as it unwinds to copy the genetic code, are movements as delicate and lovely as the dance of any angel, and the number of these molecules that could fit on the head of a pin is myriad.

Abbot Suger of Saint-Denis, one of the greatest of the Gothic builders, hoped his cathedral would reveal to worshipers the divine harmony that reconciles all discord, and inspire in them a desire to establish that same harmony within the moral order. As I examine the computer images of the molecules of life, it occurs to me that these modern electronic artifacts accomplish the same purpose as the medieval cathedrals. They inspire reverence for the invisible harmonies—of form and function, of complexity and simplicity, of sameness and variation—that define life and determine the uniqueness of every self.

As molecular biologists unravel the riddle of the genes, researchers in psychology, behavioral neuroscience, and neurobiology, along with workers in the computer-related discipline of artificial intelligence, are gaining deeper knowledge of the biochemical basis of memory. Progress has been rapid on several fronts. First, researchers have identified certain permanent structural and chemical changes that take place in brain cells (neurons) when learning occurs. Memory involves the modification of vast networks of interconnected cells; networks have been identified that are associated with different types of learning. Finally, electronic and mathematical simulations of memory networks are clarifying how memory works in the brain, and bringing closer the day when machines will be able to learn and remember in much the same way we do.

If I remember rightly, it was back about 1963 that I first became interested in the biochemistry of memory. My curiosity was sparked by some remarkable experiments with flatworms—tiny, extremely primitive animals with rudimentary brains and nervous systems. The worms were subjected to a Pavlovian training routine, in which a pulse of strong light was followed by an electric shock. After a while, the worms cringed with the onset of the light, even before they received the shock. The little fellows had learned to anticipate the shock, and had therefore "remembered" what was coming. But what was most remarkable was this: When trained worms were cut up and fed to untrained worms, the untrained cannibals also anticipated the shock. They had apparently ingested a learned behavior with dinner!

Well, maybe. It turned out that the flatworm experiments were not easy to replicate and have since been called into question. But at the time, the implication of these and related experiments seemed clear: Memories were being stored in nerve cells in a molecular way, perhaps as a sequence of chemical units in RNA or protein molecules, in much the same way as molecular DNA stores genetic information. Only molecules, it was thought, and not complex neural networks, could have survived transfer from worm to worm through ingestion and assimilation. There was something satisfying about this idea of memory as molecules. It seemed to endow memories with a kind of material permanence (more akin to "hard copy" than to "floppy disks"). But, alas, the memory-as-molecules theory has faded from fashion, as, indeed, in my own case, memory itself is fading away.

I have reached an age on the slippery side of fifty when I am increasingly afflicted by familiar faces that have no names, forgotten appointments, unpaid bills, things misplaced. I have this scary premonition that I will wake up one morning to find that *all* of my internal "floppy disks" have been inadvertently erased. There is not much solace to be had from current theories of memory. Most contemporary research confirms that memories are stored in the brain as networks of interacting nerve cells (neurons). The effect of experience is to somehow fine tune connections between cells (synapses), creating a different "trace" of interconnected cells for each memory. These spidery webs of fine-tuned synapses seem distressingly more erasable than molecules.

But if truth be told, we still don't know much about how the human brain stores information, or how it is able to call up memories at will. Many neurologists believe that we are on the verge of substantial breakthroughs in the biochemistry of memory, but my guess is that progress will be painfully slow. There are as many as a hundred billion nerve cells in the human brain. Each cell is in communication, through a treelike array of synapses, with thousands of other cells. The possibilities of interconnection are staggeringly intricate, and the problem of understanding the connections correspondingly difficult. A memory map of the human brain will be vastly more difficult to achieve than the nucleotide sequence of the human genome.

Faced with the intractable complexity of the human brain, many memory researchers choose to work with simpler organisms. The California sea snail *Aplysia,* a creature about the size of the human hand, has been a popular candidate for investigation. The sea snail's nervous system contains only about 18,000 cells, and many of these are big enough to see with the naked eye. The snails can be trained to exhibit certain behaviors in response to stimuli, and changes in their nervous systems can be monitored with relative ease. Doubtless much can be learned by studying sea snails, but there are vast differences between *Aplysia's* primitive brain and that of humans. Do sea snails suffer embarrassment when they forget a name? Do sea snails put important papers in a safe place and then forget where they put them? Has a sea snail ever forgotten its mate's birthday? Can a creature with only 18,000 nerve cells properly be said to remember at all?

No more challenging riddle remains to be solved by science than how it is that from a lifetime of experiences we can summon up remembrances of things past—sights, sounds, tastes, smells, ideas, skills, convictions. No one but the most obdurate mind-body dualists doubt that memories are somehow physically stored in that vast electrochemical system called the brain, but exactly how and where remains uncertain. Meanwhile, I start down the slippery slope of forgetfulness. Do nerve cells die with age, not to be replaced, taking memories irreversibly with them? Or is it only the ability to process memories—to call them up from data banks—that undergoes alteration with age? A huge body of scientific literature has accumulated on the subject of memory and aging—that slow and painful erasure of the self—but the answers remain frustratingly elusive.

When at last we have a complete map of the human genetic material, and a map of memory networks in the brain, will we know the self? Will the self be defined by a list of symbols in a recipe book of one thousand volumes? Will we point to a computer data bank of biochemical information (a sequence of nucleotides, a map of synaptic connections) and say "This is who I am"? Some philosophers will have no trouble with the notion that the self can be defined biochemically. Others will hold that a biochemical definition of self is simplistic and excessively reductionist. All will agree that the human self is a thing of immense complexity and subtlety which for a long time (perhaps forever) will challenge our full understanding. But the gene maps and memory maps will be made. They will advance our knowledge of our selves.

They will help us find cures for disorders of body and mind. They will open up potential futures full of promise and danger. And no philosopher who is interested in that oldest of philosophical questions can afford to ignore these remarkable developments in science.

CHAPTER 5

Sea Squirts and Space Bugs

I have a friend, a marine biologist, who haunts the beaches, saltwater marshes, and tide pools of the New England shore, collecting gifts of the sea. Now and then she finds something special that she shares with me. One memorable day she presented me with the largest and finest sea squirt she had ever found washed ashore. Sea squirts are animals that live attached to the sea floor or underwater objects. They thrive in cold water and are not uncommon along the New England shore. Externally, their bodies are not much more than little sacks with two protruding orifices, or siphons (an "in-pipe" and an "out-pipe"). If you poke a live sea squirt its muscles will vigorously contract and expel water through both orifices; hence, the name. Varieties of sea squirts take their particular names from their shape or color—sea vases, sea grapes, sea peaches, and so on. The creature my friend brought to me in her collecting bucket was a stalked sea squirt, or sea potato. You would have thought you were looking at a kind of seaweed: a soft leathery pouch on a stem, about the size of a golf ball, certainly more plantlike than animal. But the sea squirt is a member of the phylum Chordata, my biologist friend informed me, and therein hangs a tale.

Chordates are the branch of the animal kingdom that includes the vertebrates (creatures with a backbone)—cats, dogs, fishes, frogs, and of course ourselves. So it turns out that we are more closely related to the seaweedy sea squirt than to almost anything else that might wash up on the shore. Inspired by this revelation, I got out my copy of *Five Kingdoms: An Illustrated Guide to the Phyla of Life on Earth* by Lynn Margulis and Karlene Schwartz, a delightful encyclopedic catalog of the planet's multitudinous inhabitants. Of the three hundred pages in the book, the chordates are allowed five pages, the mammals two paragraphs, the primates a line, and *Homo* a single word (a glance at the book every now and then helps keep things in perspective). And, sure enough, Margulis and Schwartz have chosen to illustrate three chordates with photographs: a salamander, a swan, and a sea squirt. How, I asked myself, does that inelegant sack on stem, that spineless kelp-skinned floppy pouch, merit the company of salamanders and swans?

Taxonomically speaking, chordates are defined by the presence of three features. One is the nerve chord in the back, which in mammals becomes the brain and spinal chord. The second is a rod of cartilage, called the notochord, which forms at the back of the primitive gut in the early embryo; in vertebrates the notochord is replaced by the backbone in the course of development. The third chordate feature is the presence at some stage in the life cycle of gill slits in the pharynx or throat; the gill slits are evidence that chordates came from the sea. All of these features you and I share with the lowly sea squirt. Adult sea squirts may seem to have little in common with salamanders, swans, or humans, but in the larval stages the resemblances are striking. The larval sea squirt looks remarkably like the tadpole of a frog, or even like an early human embryo. It has rudimentary eyes, ears, and a brain, all of which the creature loses after a brief period as a free-swimming animal. It was these tadpolelike larval characteristics which led the nineteenth-century naturalist Kowalevsky to conclude that sea squirts belong to the same branch of the animal kingdom as ourselves. That discovery, which came only a few years after the publication of Darwin's *Origin of Species*, was instrumental in bringing about acceptance of the theory of evolution.

A human embryo and a sea squirt larva are kissing cousins, remarkably alike in function and form. So what happened? How did sea squirts and humans so radically diverge on the tree of life? The creature in my friend's collecting bucket certainly didn't look like a fellow chordate. It is true that inside its leathery pouch there is a mouth of sorts, a pharynx, an esophagus, a stomach, a liver, an intestine, ovaries and testes. There is a heart, and a single nerve ganglion where there might have been a brain. But there is no backbone, no brain, no skull. Some biologists believe that a tadpoley creature similar to larval sea squirts was the earliest chordate, the ancestor of us all. If so, then the evolutionary history of the sea squirt is an example of evolution running backwards, regressing toward simplicity instead of progressing toward complexity, and the metamorphosis of a larval sea squirt into a potato-pouched adult recapitulates that downhill slide. After a brief youthful fling during which it enjoys the freedom and anatomical sophistication of a tadpole, the sea squirt settles down, attaches itself to a solid surface, and simplifies. It abandons its backbone, its sense organs, its brain. It disguises itself as a plant (an aquatic potato!), weaving and waving with the kelp. But it doesn't fool anyone. In spite of its retrogressive tendencies and primitive appearance, the sea squirt is of the same stock as the highest forms of life on earth. Sea squirts and humans are cousins under the skin.

All life forms on earth—bacteria, bread mold, house flies, cabbages, salamanders, swans, sea squirts, and humans—share a common ancestry. We are all made of the same molecular building blocks; our chemistry is identical; we share a reproductive mechanism based on DNA and RNA molecules. From the point of view of the molecular biologist, the bacterium and the blue whale are more alike than they are different; both are described by a "four letter" code of nucleotides in DNA, in much the same way that the works of Shakespeare or the Bible might be translated as a Morse code of dots and dashes, short spaces and long spaces.

By definition, life reproduces, and there are adequate reasons to expect occasional "errors" in the copying process, bits of jumbled or deleted code. According to the classical Darwinian doctrine, nature selects the "errors" that are most fit. Start with a primordial organism, a bacterium for instance, wait a few billion years, and eventually the bacterium will evolve into sea squirts and biologists. But where did the first bacterium come from? No one has yet provided a satisfactory explanation of how life got started. Two theories are popular: (1) The first organisms arrived on earth from somewhere else (which merely defers the problem of explaining the origin of life); or (2) the first living organisms happened spontaneously on the early earth.

Many experiments since the 1950s have attempted to recreate in the laboratory the chemical and environmental conditions of the early earth. In these experiments most of the molecular building blocks of life—sugars, phosphates, organic bases, amino acids—have been spontaneously produced from simple inorganic substances. But the creation of the molecules that are the essence of life—DNA, RNA, and proteins—has so far eluded researchers. To make a protein, amino acid molecules must link together in a particular sequence many hundreds of molecules long. Each link in the chain is established by the removal of a water molecule. As the chain assembles it twists into a helix, like a telephone cord, and at the same time folds into a crumpled cross-linked shape that is determined by the particular sequence of amino acids. The shape of the protein decides its role in life. DNA is another chain of simple sub-units, shaped like a spiral staircase. The side rails of the staircase are sugars and phosphates, and the treads are pairs of organic bases. In a typical DNA molecule there are billions of steps on the staircase, and it is the sequence of steps that is the genetic code. DNA is the secret to life's ability to reproduce. The molecule copies itself by unzipping down the middle, and each side strand acts as a template for reassembling its complement. Segments of DNA, with the assistance of RNA, supply the templates for building proteins. Molecules make molecules. In principle, it's all chemistry.

So why can't researchers make it happen in the lab? To construct DNA and proteins from nonliving matter requires the concentration of raw materials, an input of energy, the removal of water, something to organize the raw materials, and a catalyst—conditions that have proven notoriously difficult to achieve simultaneously in the laboratory. One especially promising line of research takes a clue from Scriptures. According to Genesis, clay was the inanimate substrate for life: "Then the Lord God formed man out of the clay of the ground and breathed into his nostrils the breath of life." (Gen. 2:7) At the time Genesis was written, clay was the premier material of artisans. Of it were made containers, tablets for writing, and effigies of animals and men. So what could have been more natural than for the Creator to do his work in the same medium? According to the author of Genesis, the Lord took up clay into his hands and molded it into the beasts of the field, the birds of the air, and the first man. If British chemist A. G. Cairns-Smith is right, the story of Genesis may not be far off the mark. Cairns-Smith is the most vigorous advocate for a theory that has gained broad support among biologists and chemists: Clay may have been the catalyst that caused life to begin on this planet.

As long ago as the late 1940s, the chemist J. D. Bernal pointed out that clay is an ideal material to facilitate the synthesis of complex organic molecules. Clay is not just muck; it is a highly ordered crystalline substance with a surface affinity for organic molecules. Clay can serve as a template for organizing organic molecules into long chains. The wetting and drying cycles of clay can transfer energy from the environment to the assembling chains. And clay can shield complex organic substances from the disorganizing influence of ultraviolet sunlight. Was clay the catalyst for the first organisms? Cairns-Smith thinks so, and he goes even further. He believes that the first self-replicating molecules may actually have been filaments of clay, only later replaced by the DNA they helped to organize.

Life makes life; it happens all the time in every cell. But how did nonlife make life? It is one of the biggest unanswered questions in biology, and so far biologists can offer little more than wild guesses. Somehow inanimate matter managed to get itself organized in an animate way. Clay may or may not hold the answer to the riddle. But it is not inconceivable that at some time in the future life will become a laboratory artifact, fabricated on a bed of clay. The author of Genesis may have had it right, after all.

Whatever our common origin, this much is certain: The sea squirt is my cousin. The hummingbird and the humpback whale are twigs on my family tree. Bacteria and viruses are kith and kin. All species of life on earth share a common chemistry and a common genetic inheritance. The evidence is strong that we are all descended from a single organism or group of organisms that appeared on the surface of the earth not long after the planet formed from space dust, four billion years ago. Those ancestral cells were probably similar to the most primitive bacteria existing today.

Most scientists accept that earth's first living organisms arose spontaneously from nonliving matter, by random alliances of molecules, perhaps in primordial seas that were rich in prebiological organic compounds and crackling with energy, or perhaps on a bed of clay on a primordial shore. A less popular theory supposes that the first self-replicating cells arrived on earth from space, alive and kicking and raring to go. Those hardy "space bugs" may have been included among the dust and gases out of which the planet formed. Or perhaps they were carried to the surface of the young planet by comets, hitchhiking on Comet Halley, or as part of the dark dust clouds that drift through space. According to this view, life is common throughout space, perhaps coextensive with the universe itself, in the form of hardy viruslike or bacterialike spores. When these "seeds of life" find a suitable planetary environment—as presumably they did on earth four billion years ago—they are capable of evolving into more complex organisms, and we are but a sideshoot on the ubiquitous cosmic tree of life.

If I have relations out there among the stars, I want to know it. All of the molecular building blocks of life are apparently out there, in the materials out of which stars and planets form. As yet, no one has demonstrated the existence of living cells in space, and most scientists find it hard to believe that even the hardiest microorganisms could survive the rigors of outer space. Carl Sagan, for example, analyzed the hazards of interstellar travel for unprotected microorganisms and pronounced them impossibly severe. In particular, it seemed unlikely that microorganisms could survive the near-perfect vacuum of space, the low temperatures, and the dangerous flux of ultraviolet radiation from surrounding stars.

But maybe we have underestimated the resilience of life. Single-celled "space bugs" may be tougher than we think. Peter Weber and J. Mayo Greenberg of the University of Leiden in the Netherlands have performed a series of laboratory experiments to test whether microorganisms could survive as drifters in space. They subjected bacteria spores (*Bacillus subtilis*) to conditions of high vacuum, low temperature, and ultraviolet radiation similar to those that exist in space, and the survival rates of the spores were measured. The experiments made it clear that ultraviolet radiation is the greatest threat to cosmic spores. Surprisingly, the low temperatures of space give bacteria a greater resistance to radiation than they would have at ordinary temperatures. Weber and Greenberg estimate that a typical survival time for a naked spore adrift in space is of the order of hundreds of years; not a bad span of life, but much too short a time for a spore to complete a journey between stars.

But space-traveling bacteria might be protected from ultraviolet radiation. If the spores travel within the dark dust clouds that are known to drift through space, they would be shielded by the cloud from some of the deadly radiation. Further, in such clouds the spores could become coated with a thin mantle of ices. The Dutch experiments showed that an icy armor significantly increases the survival rate of bacteria. When both factors are taken into account, the survival time of spores in space increases to millions or tens of millions of years, more than enough time to make the drifting passage from one star system to another.

So some bacteria, at least, are likely to be hardy space sailors, capable of withstanding celestial tempest and calm. If our ancestral cells had their origin in interstellar dust clouds, or on a planet somewhere far across the galaxy, it is not impossible that their progeny made their way across oceans of space to colonize the island earth. We may yet discover that the dark nebulas—Orion's Great Nebula, for example, barely visible to the unaided eye—are arks, teeming with living cells, drifting among the stars and seeding planets. If this is so, then "the force that through the green fuse drives the flower" (to use poet Dylan Thomas's felicitous phrase) infuses the entire galaxy, perhaps even the universe, and we are a clan of creatures with kith and kin upon the planets of a billion stars.

CHAPTER 6

Sex, Sleep, Death

Some years ago, humorists James Thurber and E. B. White wrote a book called *Is Sex Necessary?* The question was not altogether frivolous and any biologist can tell you the answer: No. Moreover, sex is a terribly inefficient way to go about the business of reproduction, fraught with dangers, blind alleys, and wasted resources. Humans are so preoccupied with sex that we tend to overlook the fact that life would be much simpler without it. I'm not talking about abstinence, but about asexual methods of reproduction—cloning, sending out shoots, or parthenogenesis (reproduction by means of unfertilized eggs, seeds, or spores). What a lot of energy we waste, as a species, thinking about sex, talking about it, and doing it. And apparently it's not much different for the birds and bees, as even Thurber and White were perceptive enough to realize. Given all the fuss and bother, biologists wonder why sex evolved at all, and what sort of evolutionary pressures maintain it.

Some of us learned in school that amoebas are the only creatures that reproduce without sex, by simply splitting down the middle, after first preparing a copy of the DNA for each half. In fact, many higher-order animals and plants reproduce asexually.

Asexual species are especially common among the insects. In 1932 an all-female variety of fish was found in the Gulf of Mexico, a discovery that extended asexuality to the vertebrates. There are many species of lizards that make do with one gender. In principle, there is nothing that requires even humans to come in two varieties. "I love the idea of there being two sexes, don't you?" says the overbearing woman to her doubtful male companion in a famous Thurber cartoon. But as much as we might "love the idea" of two sexes, our opinion has little status as a biological imperative.

So why sex? For a long time the standard story among evolutionists was that sexual reproduction increases variability among offspring by mixing the genes of two parents. Variation helps populations adapt quickly to new competitors or predators and to changes in the physical environment. In this traditional view, sex speeds up evolution: In a world ruled by natural selection, the species that stands still is lost—or so it seemed according to Darwinian dogma. But, in fact, many biologists question whether sexual selection confers any clear advantage in the struggle to survive. Asexual species of "higher life" seem to survive quite nicely, and with as much genetic variability as sexual species. Consider the dandelion, an asexually reproducing plant: No one with a lawn doubts the dandelion's adaptive success. The same can be said for certain asexual weevils that show more variability than their sexual cousins, and more success in adapting to new environments. Further, computer simulations of evolving populations show that a sexual "mixing" of genes can have little or no effect on rates of evolution.

The questionable advantages of sex must be measured against the "cost" of sexual reproduction. Sexual organisms must go to the bother of finding and courting a mate, a chancy business even for creatures less finicky than ourselves. Sexual activity involves increased risk of predation and contagious disease. For males that swim or fly, a penis can be a bit of a drag, aerodynamically speaking. Females, especially, pay a high price for sex:

They sacrifice one half of their genetic legacy to a male who typically offers a negligible investment in parenthood. By contrast, the female aphids that infect the cabbages in your garden do very well without males. They produce daughter aphids in astonishing numbers with no bother at all, parthenogenically. Sometimes granddaughter aphids begin to develop while the daughters are still forming inside the mother's abdomen—aphids inside of aphids inside of aphids. It is all rather simple and bad news for the cabbages. From a Darwinian point of view, the asexual aphids seem wonderfully fit.

Sex and reproduction are two different things. Reproduction appeared with the first organisms, indeed with the first self-replicating molecules. One thing shared by all living cells is the continuous replication of DNA, the molecules that contain the genetic information. Perhaps organisms reproduce because they never stop making DNA and the other large molecules of life (RNA and proteins); when a bacterium gets too big it simply divides, each offspring taking a complete copy of the genes and half of everything else. Two individuals have replaced one. That's reproduction, and it appears to be a biological imperative. Sex, on the other hand, is gravy. Sex is icing on the cake. Life, reproduction, and even evolution could have happened without it. We could all divide like amoebas, spread spores like fungi, or bud like tree cuttings. But the fact remains that the overwhelming majority of the millions of species on earth do reproduce sexually, and as far as I can tell no one knows why. Every biologist who has considered the origin and persistence of sex seems to have a different theory, a sure sign that no theory is particularly viable. Sex is a Darwinian embarrassment, a bit of evolutionary monkey business that no one has yet satisfactorily explained.

Perhaps sex is the legacy of a series of useful evolutionary accidents that took place in bacteria billions of years ago, enduring because it became inextricably bundled up with the machinery of reproduction. Or perhaps the genes that confer sex upon an organism endure because, like all genes, it is simply their business to endure. Perhaps, after all, sex confers some yet unperceived advantage in a dangerous world, a way for the organism to stay one step ahead of the competition. Perhaps without a constant "stirring" of the genes we would be more susceptible to parasites, viruses, and other contagions. Perhaps sex is merely a fluke, a lucky or unlucky role of the dice that became embedded in evolution. Dandelions and cabbage aphids live in blissful ignorance of the raptures that elate and confound sexual creatures such as ourselves, and they do quite nicely in the struggle to survive.

Maybe James Thurber and E. B. White had it right. According to those tongue-in-cheek humorist-philosophers, males and females have always sought, by one means or another, to be together rather than apart. At first they were together by the simple expedient of being unicellular. Later, in the course of evolution, the cell separated, "for reasons which are not clear even today, although there is considerable talk." The two halves of the original cell have been searching for an appropriate other half ever since.

The title jumped off the "new book" shelf at the college library, a volume called *Why We Sleep*. Not the perennially interesting question—why do we sleep?—but the declarative promise of an answer—why we sleep. And the answer was promised with some authority; the volume bore the imprint of Oxford University Press.

But the promise was a sham. By the third sentence of the book, the author, sleep researcher James Horne, confessed bluntly: "Of course, I do not have the answer to why we sleep, as too much is still unknown." One-third of our lives is spent sleeping and no one knows why. Like sex, sleep is a gift of the genes with no discernible purpose. Horne summed up the situation this way: "Despite 50 years of research, all we can conclude about the function of sleep is that it overcomes sleepiness, and the only reliable finding from sleep deprivation experiments is that sleep loss makes us sleepy." Two centuries ago Samuel Johnson said the same thing in more somnolent language: "No searcher has yet found either the efficient or final cause (of sleep); or can tell by what power the mind and body are thus chained down in irresistible stupefaction; or what benefits the animal receives from this alternate suspension of its active powers."

So once again we are left in the dark. Pursuing the mysteries of sleep, the mountain of science has labored mightily and brought forth a mouse. Nay, not just a mouse, but thousands of mice, rats, puppies, chimpanzees, dolphins, and drowsy undergraduates, all allowed to sleep or kept awake, and watched, watched by eager researchers keen to discover why we sleep, to no avail. There is no scarcity of theories: sleep restores the body; sleep restores the brain; sleep conserves energy; sleep occupies unproductive time; sleep passes the scary hours of darkness; sleep is for dreaming. No theory has yet found convincing experimental support. I asked the teenager at our house why teens sleep so much; he said, "Ya don't havta think," and that theory is probably as good as any other.

Sleep research usually takes one of two forms. In the first, the sleeper's brain waves, brain temperature, eye movements, and muscle tone (among other functions) are monitored electrically. In the second, the subject is deprived of sleep to see what happens. The first line of research has demonstrated that normal sleep occurs in cycles and stages, usually four cycles a night and at several levels of oblivion defined by changes in the brain's electrical activity. In each cycle there is an interval of sleep when the eyes twitch under closed lids, called REM (rapid eye movement) sleep. REM sleep is the darling of sleep researchers, probably because it accompanies dreams. But what is the brain doing in REM sleep? And why do we dream? Rats have REM sleep; do they dream too? Some dolphins sleep with half of their brain at a time; does the sleeping half dream while the other half is awake? The questions are endless, but answers are few.

Sleep deprivation experiments have proved equally unfruitful.

Pity the poor rats forced to run on motorized treadmills; if they nod off to sleep they tumble head over heels. One classic piece of sleep-deprivation apparatus uses a rotating paddle to push into water the hapless rat that dares to doze. Sleep-deprived rats die within a week or two, but careful autopsies have not made clear the cause of death. Maybe the little fellows expire just to escape the rigors of the experiment. Humans show no serious impairments after a week of wakefulness. A seventeen-year-old schoolboy named Randy Gardner set the record by staying awake for eleven days; toward the end of the experiment he turned into a bit of a zombie, but quickly recovered after a long night's sleep.

Lab shelves sag beneath volumes of data, yet no one has discerned that sleep has any clear biological function. What evolutionary pressure selected this curious behavior that forces us to spend a third of our lives unconscious? Sleeping animals are more vulnerable to predation. They have less time to search for food, to eat, to find mates, to procreate, to feed their young. As Victorian parents told their children, sleepyheads fall behind—in life and in evolution. Sleep researcher Allan Rechtschaffen asks, "How could natural selection with its irrevocable logic have 'permitted' the animal kingdom to pay the price of sleep for no good reason?" Sleep is so apparently maladaptive that it is hard to understand why some other condition did not evolve to satisfy whatever need it is that sleep satisfies. Unless we have missed something, says Rechtschaffen, sleep is the biggest mistake evolution ever made. Unless, of course, you count sex.

The breakthrough that will solve the riddle is yet to come. In the meantime, we fall happily into bed at night to knit our raveled sleeves of care. The worries of the day and terrors of the wakeful night are relieved by the balm of sleep. Like the kid said, "Ya don't havta think."

The little deaths of sex and sleep are nothing compared to the great sleep that rounds our lives. Why do we die? I'm not talking about death by accident, murder, war, or disease, but the inevitable senescence that comes to us all, the catastrophic decline into old age and death that no amount of care, wealth, or connivance can delay. A lucky mayfly might survive for as long as four weeks, a turtle can live for one hundred fifty years, and a human being for a century, but when your number comes, the time is up.

Why aren't we immortal? There is at least one good reason to wonder why we live for so short a time. Evolution should favor long lifespans. The longer an animal lives, the more offspring it is likely to produce (assuming no decline in reproductive capacity), and therefore the greater the chance that its genes will spread throughout a population. In Darwinian terms, the immortal reproductively-active organism should be the fittest of all.

But maybe it's not that simple. Within most animal populations, predators drastically reduce the number of survivors before old age takes its toll. Among certain wild birds, for example, only a tiny fraction of the population survives until old age. A gene that causes senescence in birds will not be strongly selected against because the number of birds that reach old age is negligible. The situation with humans is rather different. In the developed countries, especially, human beings are increasingly likely to die of old age rather than by violence or premature disease. In the lingo of the biologist, the survival curve for humans is becoming ever more "rectangular": The percentage of survivors remains fairly constant with age until about age seventy and then plummets precipitously. But what causes the rapid decline at the proverbial three score and ten? Is the aging process triggered by genes? Or do cells in the body simply wear out by accumulating an unsupportable number of defects or waste products? Olivia Pereira-Smith of Baylor College of Medicine in Houston has reported data suggesting that genes do indeed cause aging and death. Pereira-Smith and her colleagues studied laboratory cultures of human cells (colonies of cells grown in a nutrient medium). Normal cell cultures become senescent and die after a certain fixed number of doublings. But gene mutations can lead to exceptions. Certain immortalized cell lines—cancerous cells, for example—will continue to divide forever. By performing hybridization experiments on twenty-six immortalized cell lines, the Baylor group amassed evidence to suggest that as few as four genes might be responsible for senescence in normal cells.

It's a long way from studying cell cultures in laboratory flasks to understanding entire organisms. Nevertheless, the possibility that old age and death is triggered by genes inspires a bit of wide-eyed speculation. Biologists have acquired the ability to modify genes. Might it be possible someday to engineer a strain of humans who are not programmed to die? Is immortality an option, not for us but for some future race—*Homo aeternus*? Senescence in humans is a complicated mix of subtle and obvious changes, none of which scientists yet fully understand. But if aging and death are programmed by genes, then I wouldn't bet against the possibility that extravagantly long lifetimes might someday be engineered. Sooner or later geneticists will tinker with the biological clock that ticks inexorably in every cell, and maybe, just maybe, postpone the alarm that announces decay and death.

The personal and social implications of immortality are staggering. If overpopulation is already a problem, what will happen in a world where individual human beings can live forever, assuming, of course, that they stay out of the way of germs, bullets, and speeding automobiles? And would we want to live forever if we had the choice? Do we really envy Methuselah? Can you imagine a love affair lasting nine hundred years? Or nine hundred years of television reruns? The Hyperboreans of Greek myth lived for a thousand years, free of ills, in a land of eternal sunshine beyond the north wind; they leaped into the sea like lemmings to escape boredom. It is scary to contemplate what immortality might mean for the human species. I asked a friend if he would want to discover the Fountain of Youth; no, he said, but he wouldn't mind discovering the Fountain of Middle Age. For myself, I suspect that longer lifetimes would bring more grief than bliss. Natural selection had millions of years to perfect the cellular apparatus of life, presumably to the advantage of our species. Senescence, like sex and sleep, is a sometime joyful, sometime sorrowful mystery, but I doubt if the Ponce de Leóns of genetic science will do much to improve upon Mother Nature's plan.

CHAPTER 7

The Birds and the Bees

In the days before television replaced nature in the lives of children, parents told their offspring about the birds and the bees. Or so I've heard. No one in the house I grew up in ever mentioned birds or bees (or human sex either, for that matter). But somewhere in the family library I found *The Life of the Bee* by Maurice Maeterlinck, famed Belgian man of letters, naturalist, and beekeeper. Maeterlinck's book, first published in this country in 1901, was a popular classic of natural history. Countless kids, including me, found within its pages all the sex education our innocent minds could usefully absorb.

Consider this typical passage from the chapter on the queen bee's nuptial flight: "Around the virgin queen, and dwelling with her in the hive, are hundreds of exuberant males, forever drunk on honey; the sole reason for their existence being one act of love." Even now, in these more sexually explicit times, Maeterlinck's gushing prose makes blush whatever cheek of innocence we still can turn to the central mystery of sex. What Hollywood scriptwriter ever penned a steamier copulatory scene than this: "She, drunk with her wings, obeying the magnificent law of the race that chooses her lover, and enacts that the strongest alone shall attain her in the solitude of the ether, rises still; and, for the first time in her life, the blue morning air rushes into her stigmata, singing its song, like the blood of heaven; . . . she summons her wings for one final effort; and now the chosen of incomprehensible forces has reached her, has seized her, and bounding aloft with united impetus, the ascending spiral of their intertwined flight whirls for one second in the hostile madness of love."

Victorian naturalists drew more honey from the sex lives of birds and bees than any bee ever drew from a blossom, and if they projected onto winged creatures something of their own libidos—well, it has been a commonplace since Aesop to endow animals with human traits. Modern naturalists report their observations in language less fervid than Maeterlinck's, but they still, perhaps, project. Examples from the current scientific literature are not hard to find.

I read with interest certain observations of pied flycatchers, small black-and-white birds that have been favorite subjects for those who study animal behaviors, reported in *The American Naturalist* by Norwegian ethologists Eivin Roskaft, Jan Ove Gjershaug, and Torbjorn Jarvi. Their article is entitled "Marriage Entrapment by 'Solitary Mothers': A Study on Male Deception by Female Pied Flycatchers." A more evident example of projection can hardly be imagined.

Pied flycatchers are polygynous (a single male has several mates). One female in the menage is primary, and only she gets help from the male in raising the young. Secondary females must shift for themselves, and their reproductive success can be (according to our intrepid researchers) as much as 55 percent lower than that of their primary sisters, a circumstance that might be expected to evoke, on good Darwinian principles, some compensatory behavior. The Norwegian ethologists looked for such behavior. Because of the low frequency of secondary females in their study area, they experimentally "widowed" a number of primary females, thereby creating "de facto" secondaries. Twenty males were removed from their territories after their mates had laid eggs. Of the widowed females, seventeen were visited by neighboring males. Six of the females were observed soliciting copulation out of season—something never observed of primary mothers. Three females actually copulated, and one was successful in getting her new mate to adopt her brood of nestlings. Roskaft and colleagues conclude: "Our study shows that, by soliciting copulations, widowed females may have led the new males to believe that they were fathers of their broods. The new males were thus fooled or trapped, even though the females had already laid their eggs and were thus temporarily infertile." And "thus, it seems evident that some pied flycatcher . . . males can be bluffed by 'widowed' females into adopting their young."

The attribution of such deliberate cunning to female pied flycatchers sounds more like Maeterlinck than modern science. We can easily imagine a Victorian father using the story of pied flycatchers to warn his son against conniving women who might "fool" or "trap" him into matrimony. The respected journal *Science* reinforced this interpretation; its report on the pied flycatcher experiment was titled—with breathtaking generality—"How Females Entrap Males."

Given the fact that in this particular experiment only one of twenty "widowed" flycatchers found a male willing to take on the permanent responsibilities of fatherhood, it may be stretching the point to conclude general perfidy on the part of the female of the species. Is one permitted to wonder if breaking up twenty happy households in the name of science is not a greater perfidy than whatever poor, desperate "bluff" the female flycatcher employs to increase the sustenance of her brood? The voluptuous language of Victorian naturalists may make us blush, but at least they drew their morals explicitly. Maeterlinck never doubted our affinity with the bee, and had no compunction about using words like "marriage," "mother," "fool," and "entrap" to describe bee behavior. Modern naturalists pretend greater objectivity, but do the same thing.

In the same vein, here's one for the habitués of the singles bars.

Looking for the perfect mate? Or just a one-night stand? What defines a good pickup bar? A choice location? A standout crowd? Who's pulling the strings—I mean *really* pulling the strings—that control the pickup dynamic? Is that group of gals at the bar just waiting for you to make your move, or have they already made it for you? Are those guys at the corner table seriously looking for love, or are they just playing games among themselves? Is any of this really necessary? Listen and learn from the swallow-tailed manakin.

Or, less specifically, consider the lek-breeders.

Lekking is a system of mating practiced by certain birds, frogs, bats, and insects. Males gather in groups called leks to make themselves available to females. The female visits the lek, observes the males on display, and makes her choice. After copulation with the favored male, the female goes off to rear the eventual offspring on her own. Biologists have long wondered what evolutionary pressures gave rise to lekking in animals as diverse as birds of paradise and dragonflies. Until recently, two theories have dominated discussions of lekking behavior, both originally offered by Jack Bradbury of the University of California at San Diego and his associates. The first theory is called the female-preference model. In this view, males gain nothing from gathering in one place for their courtship displays. It is the female who prefers clustered males, presumably because it makes the business of selecting a desirable mate more efficient. Having many potential partners gathered in one place makes for easy comparison shopping; the female quickly discerns the dominant male and makes off with the fittest genes. The second traditional explanation of lekking behavior is called the hotspot model. Males gather at those special places near a food source or desirable nesting site where they are likely to encounter females. When a female arrives at the hotspot, the males hover, somersault, sing, display plumage, or do whatever it is that demonstrates their genetic suitability as a mate. The female then makes her choice from among the available males.

In both traditional theories, it is female behavior that determines the formation of the lek and the mating success of the male. Bruce Beehler and Mercedes Foster, two animal researchers from the Smithsonian Institution, challenge the conventional wisdom by emphasizing the male contribution to lekking. They find no compelling field data to suggest that males cluster because females prefer groups. As for leks making the female's choice more efficient, the two researchers believe the opposite may be true. Male competition within the lek may in fact disrupt mating activities and limit freedom of choice for the female. Instead, the issue of which male will mate with her may have already been decided among the males before the female shows up at the lek. In one study of the lesser bird of paradise, a single male in the lek performed 24 of 25 copulations. Similar dominance has been observed among insects. The overwhelming success of certain males within a lek was previously ascribed to unanimity of female choice. Beehler and Foster ask how female preference could be so unanimous among males that show—to the human eye, at least—so little difference in looks or behavior. As evidence for their view, the two naturalists point to the swallow-tailed manakins.

Before any female shows up on the scene, male manakins (small, brightly colored tropical birds) gather in the lek to engage in competitive behavior. One dominant and one subordinate male emerge from these rituals. When a female arrives, these two perform various displays, and then the subordinate male retires to let the dominant male get on with his business. Female choice is nonexistent. In place of female-preference and hotspot theories, the Smithsonian researchers offer what they call the hotshot theory of lekking behavior. According to this view, certain males, for one reason or another, are more successful at attracting mates. Other, less successful males gather around these hotshots in the expectation that they will have access to more females than if they displayed alone. But why does a hotshot tolerate hangers-on who are potential rivals? Beehler and Foster believe benefits might accrue from decreased chances of predation (safety in numbers), and from the possibility that a crowd of males will attract more females to the courtship arena. As long as the hotshot maintains control of the lek, the larger number of visiting females is all to his benefit.

Somehow it all sounds familiar. The guys go where the gals are. The gals go where the guys are. The gals are looking for Mr. Right. The guys strut and bluster, and play macho games among themselves. No one really knows what's going on behind all the posturing and pairing. In fact, lekking among the birds, bats, frogs, and dragonflies seems nearly as complicated as lekking among humans.

So how *do* I make myself more attractive to the opposite sex? Should I invest in snappy clothes or in posh real estate? Should I dress in bright colors or wear muted earth tones? Should I sign up for a bodybuilding course at the local health club or cultivate a lean and hungry look? And what can the birds and the bees (and the scientists who study them) teach me about all of this? Maybe nothing. But we have long been inclined to see glimmers of ourselves among the lower orders of animals, and glimmers of them in us. We have enough in common with birds and bees to want to search their behaviors for instruction about our own. And so it is that I attend to articles in the scientific literature on the sexual behavior of pied flycatchers, swallow-tailed manakins, and Australian bowerbirds.

The bowerbirds of Australia and New Guinea have one of the most elaborate mating behaviors in the animal world. The males of the species build wonderfully decorated bowers for the purpose of attracting females, and as sites for copulation. The bowers are built with sticks on cleared ground. The floor of the bower is covered with bright straw and adorned with such baubles as snail shells, bright pebbles, iridescent feathers, insect parts, and bits of bone. Objects purloined from human abodes are often used to enhance the bower: bottle caps, thimbles, coins, clothespins, and the like.

Zoologist Gerald Borgia used automatic cameras to keep close tabs on a colony of bowerbirds, recording even their most intimate moments. His conclusion: Female bowerbirds actively choose mates on the basis of the quality of their architectural displays. Borgia and his colleagues went so far as to remove the decorations from certain bowers. The owner's success with females plummeted. Presumably, sexual behavior among bowerbirds is genetically programmed so that females select mates with "good genes." Says Borgia: "A female that chooses an older, established male with a well-built, well-decorated bower and a refined courtship call has evidence that her prospective mate not only has been able to survive to a relatively old age, but also has been able to do it while learning to build and maintain a high-quality bower under the rigors of male competition." We older, established human males with well-built bowers and a repertoire of sophisticated "sweet talk" can't help but feel a certain sympathetic comradeship with bowerbirds.

On the other hand, we may be less consoled by pied flycatchers. In one study, researchers chose male flycatchers at random and forced them to occupy nesting sites in territories of varying quality. It turned out that the age, size, plumage color, or song repertoire of the males had no influence on female choice. The single most important criterion was the quality of the occupied territory. Female flycatchers wanted a home safe from predators and close to favored feeding sites, and chose any old mate to get it.

By selectively breeding ladybugs over many generations, researchers have demonstrated that female preference for male ladybugs (gentlemenbugs?) of a certain coloring is controlled by one or a few genes. Almost certainly, the sexual behaviors of flycatchers, manakins, and bowerbirds are determined by genes. If the sex lives of birds and insects is held in thrall by tiny segments of DNA, then what of ourselves? Are the mating behaviors of *Homo sapiens* arbitrary social conventions, or are they stitched by evolution into our genes?

It has now been more than a decade since the publication of Edward O. Wilson's *Sociobiology: The New Synthesis*, a magisterial analysis of the genetic basis for animal behavior. The book evoked a lively controversy among anthropologists and sociologists. What exasperated many of Wilson's colleagues was his claim (in the last chapter of the book) that human social behavior, like that of the birds and bees, is genetically constrained. According to Wilson, such things as sexual behavior, aggression, and social stratification are at least partly controlled by genes. That claim was ideologically and scientifically unacceptable to many people, ideologically because it seems to make us moral hostages to evolution, scientifically because conclusive evidence for a genetic basis of human behaviors is almost impossible to obtain (selective breeding of humans over many generations is clearly out of the question).

Perhaps, as the sociobiologists claim, we share enough of our genetic inheritance with the lower orders of life to find in their behaviors instructive parallels to our own; in this, sociobiologists are the new Maeterlincks. Or perhaps, because of our bigger brains, human sexual behavior has broken entirely free from the bonds of instinct; in which case it is just as well that children no longer look to the birds and bees for sex education. As one who was brought up on Maurice Maeterlinck's gushy evocations of the raptures of bee sex, I am inclined to believe that genes have a not insignificant grip on our sexual behaviors. At the very least, I'm willing to learn from bowerbirds, flycatchers, and manakins.

CHAPTER 8

The Grisly Folk

Love and death: the two great themes of human literature and Woody Allen films. If birds and bees have something to tell us about our innate propensities for sex, a consideration of our earliest human ancestors reveals a grim propensity for violence.

Pity the poor Neanderthals, who had the misfortune to be discovered at about the time Darwin was evoking the outrage of his contemporaries by suggesting that humans, apes, and gorillas have a common ancestry. The fossilized bones of Neanderthals were first excavated and studied in the mid-nineteenth century. The bones were undeniably human, but distinctly different from those of modern men and women. The stocky limbs and heavy, slanted brows suggested a gorillalike ancestor that no proper Victorian welcomed to the human family tree.

Until recently, Neanderthals were considered dull-witted, brutish aberrations of evolution, blessedly rendered extinct about thirty-five thousand years ago by the rise of anatomically modern humans. Even as late as my own childhood, the implicitly racist story of the displacement of the Neanderthals by tall, fair-skinned Cro-Magnons (our own direct ancestors) had all of the elements of an edifying moral tale—the triumph of European cleverness and blond good looks over all that was barbarous and base. In his widely read book, *The Outline of History*, published in 1920, H. G. Wells promoted the view that a dim species remembrance of Neanderthals may survive in our folklore stories of ogres. He assumed that the first modern humans did not interbreed with Neanderthals, and attributed this separateness to the Neanderthal's "extreme hairiness," "ugliness," and "repulsive strangeness." In a short story, Wells referred to Neanderthals as "the grisly folk." In Wells's version of prehistory, which for a long time was shared by scientists, the triumph of modern humans over Neanderthals was the triumph of reason, imagination, and lofty moral vision over ugliness, stupidity, and amorality.

The novelist William Golding, best known as author of *Lord of the Flies*, was one of the first to suspect that our brutish image of Neanderthals was actually a projection of unattractive characteristics that we found within ourselves. In his 1955 novel, *The Inheritors*, Golding turned the story of Neanderthals and Cro-Magnons on its head. Golding's Neanderthals live in a state of childlike innocence, possessed of wonder and imagination. They do not willfully kill other animals. They are sexually restrained, and charmingly uninhibited about their nakedness. Into their Edenlike existence come the violent and cannibalistic Cro-Magnons. The new folk revere a witch doctor with an antlered mask. They are adulterous and engage in orgies. The gentle Neanderthals are no match for the craftiness and cunning of the new arrivals. Blood flows. Except for a single child, Golding's happy band of Neanderthals is exterminated, and the tougher, more aggressive Cro-Magnons inherit the earth.

Golding's revisionist story has its parallel in science. In recent decades anthropologists have been taking a fresh look at the fossil evidence, and now concede that Neanderthals were creative, imaginative people that greatly extended the regions of the world occupied by humans. They practiced highly developed skills for making tools, clothing, and shelter, cared for the aged and handicapped, and buried their dead, sometimes including masses of flowers in the burial. Neanderthals were apparently not so "grisly" after all. Also, it now seems clear that modern humans did not evolve *from* Neanderthals, but that Neanderthals and Cro-Magnons are parallel branches of the human family tree. Anatomically modern Cro-Magnon fossil bones unearthed in Israel are nearly one hundred thousand years old, contemporaneous with the heyday of Neanderthals. Apparently the two subspecies of humankind are almost equally ancient, and shared the planet without interbreeding for at least fifty thousand years. It might even be appropriate to consider that Neanderthals and Cro-Magnons are separate species.

A host of questions remains to be answered. What was the relationship between the two human populations? Did they share common habitats? Did they compete for resources? And what were the features that enabled modern humans to survive and become "the inheritors" even as the Neanderthals faded away? Was it mental capacity, language, and inventiveness that gave Cro-Magnons the advantage? Or was it aggressiveness, rapacity, and a shrewd instinct for self-advantage? And finally, the old, dark question is raised again with new force: Were our Cro-Magnon ancestors implicated in the final disappearance of the Neanderthals? Many paleontologists believe that the expansive rise of modern humans, armed with new killing technologies, was the cause of the mass animal extinctions that occurred near the end of the last ice age, when mammoths, mastodons, saber-toothed cats, and a host of other creatures vanished from the earth. Could it be that Neanderthals—a parallel branch of the human family tree—were also victims of the technologically sophisticated inheritors? If so, then William Golding was right, and the "grisly folk" are really a carefully suppressed aspect of ourselves.

The extermination of Neanderthals was not the first chapter in humankind's history that is steeped in blood. From Swartkrans cave in the Transvaal region of South Africa comes news of what may have been the earliest known use of fire—and the hint of another grim episode of competition and extinction.

Swartkrans has long been a premier source for fossil evidence of early hominids (the family of primates of which *Homo sapiens* is the only survivor). The cave has three levels of fossil-bearing rock, distinguished from one another by episodes of erosion. Animal fossils and stone tools throughout the levels suggest that the deposits are between 1 and 1.8 million years old. Two different species of hominids apparently inhabited the cave or were carried to it as prey: *Homo erectus*, our direct ancestor, and *Australopithecus robustus*, another branch of the hominid family tree that became extinct about 1 million years ago. The ancestry of these two species of hominids, and the relationship between them in places where they coexisted, is a matter of intense debate among anthropologists.

One or both of these creatures may have used fire at Swartkrans. Archeologists C. K. Brain and A. Sillen, of the Transvaal Museum and University of Cape Town, report that animal bones from the most recent cave deposits appear to have been altered by fire. To ascertain what kind of changes might have been induced by fire, Brain and Sillen subjected the bones of a fresh hartebeest to the temperatures expected from a campfire. A microscopic examination of the freshly charred bones and the fossil bones showed similar changes in structural details. Chemical analysis of fossil bones from the most recent deposits yielded quantities of carbon consistent with charring; bone from older levels of the cave contained less carbon. The archeologists conclude that fire was not used by the earliest inhabitants of the cave, and therefore date the "discovery of fire" (at Swartkrans) to sometime between 1 and 1.5 million years before the present.

Charred bones do not necessarily mean that fire was used for cooking. The bones may have otherwise found their way into fires used for protection or for warmth (or, indeed, the bones might have been charred by fires of a natural origin). Nor does the evidence from Swartkrans indicate whether it was *Homo* or *Australopithecus* or both that used fire. A single specimen of charred bones of *Australopithecus* hints at something less than amicable relations between the species. As far as I know, humans are the only fire-using animal. I've heard it said that cheetahs and hawks will position themselves to attack animals fleeing from naturally occurring fires, and that tarsiers (little primates of Southeast Asia) will sneak into villages at night and grab hot coals from fires, but none of this qualifies as "the discovery of fire." Whoever lived at Swartkrans may have known how to retrieve fire ignited by volcanoes or lightning, and how to keep it alive. And perhaps the consistent use of fire is as good a criterion as any other for defining that magic moment when hominids can be said to have become human. If Brains and Sillen are correct in their interpretation of the evidence, sometime about 1.5 million years ago a clever Promethean ancestor at Swartkrans figured out that glowing coals can be carried from place to place, and that fire is useful for protection, light, or warmth. It could not have taken long, once campfires were common, to discover that cooked meat tasted better than raw meat and took less effort to chew, and that the fats and protein of meat provided more energy than a vegetable diet. Possibly later came the discovery that fire could be used to harden wooden spears or as a controlled ally in hunting.

As late as 1970, conclusive evidence for the discovery of fire did not go back more than about half a million years, but even then anthropologists ascribed all sorts of cultural significance to fire's discovery. The requirement of tending a fire presumably led our ancestors to a more settled lifestyle. The hearth was a place for communal life, and therefore for new kinds of communication—dance, storytelling, and decorative and symbolic arts. In the most imaginative of these flame-lit scenarios, happy bands of early humans sat next to a fire, swapping yarns, cooing to infants, sharpening spears, sharing tidbits of roasted meat, and taking from the hissing, crackling flame, and from the smoke curling heavenward, new and elaborate ideas about life, death, sex, and immortality.

Now that the discovery of fire has been pushed back another million years or so, somewhat less edifying scenes can be imagined. How about this one? A little band of hunters of the species *Homo erectus* come to the cooking cavern where their fire, tended by the weaker members of the band, is protected from wind and rain. On the menu at one time or another (taking my cue from the bones at Swartkrans) is antelope, zebra, warthog, baboon, and—depending on availability—an occasional *Australopithecus robustus* (from whom *Homo* may have diverged only a million years earlier), roasted to perfection, thereby hastening our smaller, less erect, tool-making cousins toward eventual extinction.

In Jean Auel's blockbuster novel *The Mammoth Hunters* the beautiful, blond Cro-Magnon Ayla muses on the coming hunt. "How could creatures as small and weak as humans challenge the huge, shaggy, tusked beast, and hope to succeed?" she asks herself. Her question evokes a familiar image: A little band of half-naked humans, shivering with ice-age cold and fear, thrusting pitiful wooden sticks at a great raging woolly mammoth. The odds seem so uneven, the foe so formidable, that our hearts and admiration go out to lovely Ayla and her brave companions. But the odds were not so uneven, after all, as Ayla knows. She has in her favor "the intelligence, experience, and cooperation of the other hunters," including handsome Jondalar and his new spear-throwing device. The mammoth herd goes down without a single human casualty, apart from a bad scare for Ayla. The same resourcefulness in killing that led *Australopithecus* and Neanderthals to extinction now brings the woolly mammoth to the brink of oblivion.

Scientists know a lot about the woolly mammoth. Several well-preserved mammoths have been recovered by Soviet scientists from Siberian permafrost, including a baby mammoth (affectionately named Dima by its discoverers). The stomach contents of the frozen mammoths even tell us about their feeding habits. More is probably known about the biology of mammoths than of any other extinct creature. Mammoths were apparently numerous throughout the last Ice Age, in Europe and Asia and western North America. Then, quite abruptly, they became extinct about eleven thousand years ago, just as the Ice Age ended. A simultaneous wave of extinctions swept many other large land mammals from the earth. The mastodon, the cave bear, the giant deer, the saber-toothed "tiger," the tanklike glyptodont, the North American camel, beavers the size of bears, and the giant ground sloth all disappeared, and with their demise the Golden Age of Mammals came to an end.

What was the cause of the extinctions? Some scientists lay the blame on a warming climate that accompanied the retreat of the great continental glaciers; among the various species of life on earth, large land animals would have been least likely to adapt successfully to the new climatic conditions. But there are problems with the climate theory of extinction. We know, for example, that the recent Ice Age is only one of many that have affected the earth over the past few million years. Why did the giant mammals survive previous interglacial periods only to fall victim to the most recent warming? Circumstantial evidence points to Ayla and her people as the agents of extinction. At several killing sites in the American southwest, finely crafted flint points have been found in conjunction with mammoth bones. Radiocarbon dating confirms the simultaneous disappearance from Arizona of the cold-adapted Harrington's mountain goat and the warm-adapted Shasta ground sloth. Climatic change would not be expected to adversely affect both species, but the disappearances of the goat and ground sloth coincide exactly with the arrival of human hunters in Arizona.

Ice Age hunters were powerfully armed with flint-tipped spears and devices that amplified the force of the arm when the spear was hurled. They were not the brave naked primitives we have sometimes imagined them to be. They were well-adapted to the tundra. They knew how to use fire as a weapon. They were skilled at setting traps. Says physiologist Jared Diamond: "A realistic painting of mammoth hunting should show warmly clad professionals calmly spearing a mammoth already crippled by a trap or ambush." It may be difficult to comprehend, but small bands of Ice Age hunters, ranging over three continents, probably possessed sufficient skills to drive several races of huge beasts into extinction, in a bath of gore that far exceeded human needs for food, clothing, and shelter.

"Hunting mammoths can be very exciting," gushes Ayla after her narrow escape from the great woolly beast. She looks at Jondalar with an aroused, blood-lust gleam in her eye. He returns her provocative glance. The steamier bits of Jean Auel's novel are about to ensue. The Golden Age of Mammals ends and the Golden Age of Humans begins.

Anthropologist Napoleon A. Chagnon, of the University of California at Santa Barbara, has studied the Yanomamo people of the Amazon for twenty-four years. What is most striking about the tribe is the ferocity of the males, who apparently revel in violence. Thirty percent of Yanomamo males die violently; almost half of the males over the age of twenty-five have participated in a killing. What do they fight over? "Reproductive resources," according to Chagnon. Or to put it more simply, women. Yanomamo killers have on the average two and a half times as many wives and three times as many children as nonkillers. Chagnon hints at a kind of Darwinian logic behind this propensity for violence. The Yanomamo male who kills his neighbors in pursuit of women increases his reproductive success, and presumably insures the continuance of his own line. If evolution is driven by selfish genes, then the Yanomamo killer is "fitter." Chagnon suggests that a connection between the procreative drive and violence may be innate and common to all humanity. "One thing you can never get enough of is sex," he says. Sex and violence go hand in hand, he seems to be saying, and the Yanomamo merely act out tendencies that civilized societies strive to hold in check.

It must be said at once that Chagnon's work is highly controversial. Not all anthropologists believe that an inborn tendency toward sex-driven violence is required to explain Yanomamo behaviors, or even toward violence in general. They point to environmental stress and competition for material resources as alternative explanations, explanations that may apply to earlier examples of internecine violence in our family history. The debate over Chagnon's interpretation of Yanomamo violence is merely a skirmish in a broader war between sociobiologists and their critics. At the heart of the sociobiology doctrine is the assumption that some human social behaviors have a genetic basis, and these include male aggressiveness, a propensity toward violence, and, of course, male and female sex roles. "It pays males to be aggressive, hasty, fickle, and undiscriminating," writes sociobiologist Edward O. Wilson. "In theory it is more profitable for females to be coy, to hold back until they can identify males with the best genes." In his book *On Human Nature*, Wilson is emphatic. "The genes hold culture on a leash," he writes. "The leash is very long, but inevitably values will be constrained in accordance with their effects on the human gene pool." In a *Scientific American* report on Chagnon's work, Wilson is quoted in support of a biological interpretation of Yanomamo violence: "I'm really curious about why people pussyfoot around the human aggression element.

Humanity has been wading in blood for as long as it's been around. If we have a strong biological predisposition toward violence, we just can't wish it away."

The debate over sociobiology has generated a huge volume of words, and more heat than light. Perhaps, as Wilson says, in rejecting the influence of genes we are wishing away an inborn tendency toward violence that is part of our biological heritage. Certainly, the history of our species is amply colored by the stain of blood. Nevertheless, in my opinion, the issue is far from settled. Nature versus nurture arguments have been around for a long time, and are not likely to go away soon. Indeed, from the strictly scientific point of view, the two sides may not be all that far apart. As philosopher of science Michael Ruse has written: "Explanations involving genes are never of the genes alone, but always of genes as they interact with the environment." Neither the sociobiologists nor their critics believe that human social behavior is fully hostage to the genes. The enormous adaptability of the human brain may have enabled human culture to escape entirely from the grim calculus of natural selection. On the other hand, there may be more Yanomamo-like behavior locked up in our biology than we care to admit, predisposing us toward the violence that brought *Australopithecus robustus*, Neanderthals, and the giant mammals to premature extinction.

CHAPTER 9

The Red Sickness of Battle

Whether or not young Yanomamo warriors go out to bash the heads of their neighbors in response to the petulant urging of their genes, they are certainly constrained to do so by cultural conventions of their society. Edward O. Wilson is correct when he says that "humanity has been wading in blood for as long as it's been around." Even such elaborate cultural conventions as the great religions—Christianity and Islam, for example—with their messages of brotherly love, have sometimes seemed more steeped in gore than in benevolence. Sociobiologists and anthropologists ponder this propensity for violence, but whether a scientific understanding of the roots of violence will help us ameliorate that tendency remains to be seen. Darwinian evolutionists tell us that nature is "red in tooth and claw," so it is perhaps a bit starry-eyed to imagine that the human species, alone of the creatures of the earth and only recently raised to the level of scientific discourse, might suddenly repudiate whatever is the contemporary equivalent of the flint-tipped stick.

In Stephen Crane's American classic *The Red Badge of Courage*, young Henry Fleming goes off to war stirred by dreams of heroic sweep and grandeur: "He had read of marches, sieges, conflicts, and had longed to see it all. His busy mind had drawn for him large pictures extravagant in color, lurid with breathless deeds." In the war to preserve the Union, Fleming would mingle in one of the great affairs of the earth. He longs, yes longs, for the nonfatal symbolic wound, the moral imprimatur for the young male, the blood-red badge of courage. The first foes Fleming encounters are a group of Confederate pickets along a river bank. One night while on guard duty he converses with one of them across the stream. After some moments of banter, the Confederate says to him, "Yank, yer a right dum good feller." That brotherly sentiment cast onto the still night air makes the young soldier momentarily regret the war and his longing for violence. By novel's end, regrets have multiplied, and Fleming rids himself at last of the "red sickness of battle," and turns with a lover's thirst to images of tranquil skies, fresh meadows, cool brooks, and peace.

The contemporary equivalent of the picket line is the surveillance satellite. The modern substitute for Stephen Crane's regiment of youths arrayed to repulse the enemy's charge is the Strategic Defense Initiative (SDI), popularly known as Star Wars, announced by President Reagan in 1983. A million years of incremental technological progress separated the stone hand ax (with which *Homo erectus* did violence to his less cerebral cousins) from the Civil War musket. In the century that has passed since Stephen Crane began his career as a writer, the technology of killing experienced a greater escalation than in the previous million years. SDI is more than an escalation; it is a quantum jump into an entirely new concept of national defense. There are those who believe that SDI will increase the security of our nation, and lessen the chance of an enemy missile strike on American targets. And there are others who think the program is obscenely expensive, unworkable, and dangerous. The fortunes of SDI rise and fall with each U.S. election and with the vagaries of international politics. But the bottom line is this: Staggering sums of money are at stake, perhaps hundreds of billions of dollars for research and development, and the temptation to have a slice of such an enormous pie may prove overwhelmingly strong. The sheer magnitude of potential funding available for SDI research poses a colossal moral dilemma for the scientific and technological communities.

It is hard to get one's moral teeth into SDI. It offers no drum rolls, no bugle calls, no "brass and bombast" to call young men to war. Henry Fleming was made to feel sublime for a moment by the frenzy of the rapid, successful charge, the music of tramping feet, the sharp voices and clanking of arms of young men engaged in combat. It is rather more difficult to romanticize a war fought with SBKKVs, ERISs, and HEDIs, to mention but a few of the components of SDI, and that, I suppose, is good. But neither does SDI offer the opportunity for friendly banter between foes across a stream, the "right dum good feller" token of shared humanity that carries within it the seed of doubt about the efficacy of violence. What sort of novel might Stephen Crane have made out of the high-tech vocabulary of SDI:

- SBKKV: space-based kinetic kill vehicle. Rockets carrying nonnuclear homing vehicles that will seek out and collide with enemy missiles in the launch phase of their flight. Several rockets would be based on each of thousands of platforms "parked" in low-earth orbit.

- Pop-up X-ray laser. These relatively light weapons will be "popped up" into space, presumably from submarines, to counter an enemy missile launch. They will convert a portion of the energy of a nuclear explosion into a powerful beam of X-rays directed at the target.

- Free electron laser. Hugely powerful ground-based lasers that will direct beams of radiation off orbiting mirrors onto enemy missiles.

- ERIS: Exoatmospheric reentry-vehicle intercept system. Long-range ground-based rockets that will intercept enemy missiles while they are still in space.

- Space-based particle beam. This device would emit a beam of subatomic particles and cause them to collide with targets. The nature of the reaction between beam and target will enable sensors and computers to discriminate between warheads and decoys. Hundreds of orbiting particle beams will be needed, each powered by a nuclear reactor powerful enough to supply a small city with electricity.

- HEDI: High-endoatmospheric defense interceptor. Ground-launched rocket designed to intercept enemy missiles in the last reentry stage of their flight.

In the event of a full-scale nuclear attack, all of these systems and more would be brought instantly into play in a battle observed by satellites and controlled by computers. In such a war, there will be no badges of courage. No youths will exchange dreams of heroism for dreams of peace. Space will blaze with bombs and beams. It will all be over in an hour.

"Science will make the Kraals invincible."

That line was spoken on "Doctor Who," the British science fiction serial that has been aired on American public television. The speaker was Styggron, the leader of the Kraals, an alien race at war with Earth. The idea that science can confer invincibility is a stock theme of science fiction. It is also the rationalization behind SDI. Proponents of SDI are determined to turn science fiction into science fact. Building the system will require major breakthroughs in materials, lasers, electronics, power plant design, and computer software development. Not since the Manhattan Project of the Second World War, the government-sponsored program to build the atomic bomb, has the scientific community been asked to turn so much inventiveness and creativity toward the technology of war, or been asked to take a greater step into the unknown. Huge amounts of research money are at stake. The project is fraught with moral and political dilemmas. Not surprisingly, SDI has split the scientific community into two highly vocal camps. Many scientists believe the project is morally or politically indefensible, or simply technically impossible. Others agree with ex-President Reagan that the project will serve the cause of peace by making war "unthinkable," and they are confident that technical problems can be solved.

The technical problems implicit in SDI are formidable. For example, the heart of the system will be powerful lasers that will knock enemy missiles out of the sky at the speed of light. Batteries or solar panels are not suitable for supplying the huge amounts of energy these space-based weapons will require. It will be necessary to put into space power plants of a far greater capacity than anything previously attempted—multi-megawatt nuclear power stations to supply energy to orbiting battle platforms. If you can imagine putting a small Chernobyl nuclear reactor into the *Challenger* space shuttle and blasting the whole package safely into space, you will have a sense of the problems that must be solved.

A second technical problem is that of writing the computer programs that will operate SDI weapons. If SDI is ever used in war, computers will control every aspect of the system. Human decision-making will be too slow. Computers will identify enemy missiles and decide if their intent is hostile. Computers will make the decision to fire. Computers will assess damage. Computers will have a power over the fate of the human race unlike anything ever before imagined. It has been estimated that the required software could consist of one hundred million lines of programming code, written by hundreds or thousands of individual programmers. That is hundreds of times more code than is required for controlling a mainframe computer. Many scientists believe that writing and debugging such programs is an impossible task, that the programs would be susceptible to "viruses" implanted by a potential enemy, and that in any case the programs could only be realistically tested in a battle situation. Other scientists hope that by appropriately delegating battle decision-making to many independent computers, the system can be tested and made to work.

And there are more things to worry about: the laser weapons themselves, the communications system, and the surveillance devices—all involving new technologies.

There is no question that if SDI (or some future equivalent) goes forward dramatic breakthroughs will accrue to science, just as the Manhattan Project led to increased understanding of the atom. And there will be "spin-off" technologies—materials, electronics, lasers—of potential benefit to mankind. On the other hand, a staggering amount of money and human effort will have been spent on a system of war that many experts believe cannot work and is undesirable even if it does. Never before has the scientific community been asked to confront technical, political, and moral questions of such complexity. Perhaps the most fundamental question is whether, like the Kraals, we should rely upon science to make ourselves "invincible," with weapons whose use is declared "unthinkable."

SDI will not make us invincible, nor is nuclear war unthinkable. There are hundreds of scientists whose business it is to do nothing but think about the weapons of nuclear war—how to use them, how not to use them, how to build them, how to get rid of them, and what the consequences of their use might be. Some of these scientists believe nuclear weapons are essential to our national defense. Others are passionate advocates of disarmament. All of them are paid to think about the unthinkable.

In 1986, a group of researchers from the U.S. Forest Service,

NASA, and the Defense Department set fire to six hundred acres of the Angeles National Forest, near Los Angeles, California, so that they might better think about the effects of nuclear weapons. The scientists flew in a variety of aircraft through the smoke cloud produced by the fire to measure its volume and opaqueness to sunlight. A knowledge of these factors is necessary if one is to accurately estimate the climatic effect of fires ignited by a nuclear war. In particular, the researchers want to know if smoke produced by burning cities in the aftermath of a nuclear attack will cause a "nuclear winter."

The "nuclear winter" theory came to public notice in 1983 when an interdisciplinary group of prominent scientists announced a new assessment of the possible consequences of a nuclear exchange. Their names were Turco, Toon, Ackerman, Pollack, and Sagan—TTAPS for short. They were not primarily concerned with the estimated 1.1 billion fatalities that would result from blast, fire, and radiation in a full-scale nuclear attack (one-fifth of the population of the planet), nor with the equal number of blast-related injuries requiring medical attention. Their concern was with the three billion human beings who would not be immediate victims of the attack, including those in nations far removed from the conflict. The "nuclear winter" scenario proposed by TTAPS was horrifying (as if anything could be more horrifying than what was already known about the destructive effect of nuclear weapons). According to their carefully constructed mathematical simulations, dust raised by blast and smoke from burning cities and forests would attenuate the sun's light and heat sufficiently to plunge the Northern Hemisphere, and perhaps the entire planet, into cold and darkness for many months. The TTAPS group concluded that the extinction of a large fraction of the earth's animals, plants, and microorganisms is possible, and that the extinction of the human species cannot be excluded. (Against these estimates the terrible slaughters of Shiloh, Antietam Creek, and Chickamauga pale, and even the extinctions of *Australopithecus,* the Neanderthals, and the great ice-age mammals seem but minor perturbations in the history of life). Perhaps the most frightening part of the study was the conclusion that a nuclear winter could be initiated by the use of even a small fraction of the weapons currently available in nuclear arsenals.

The TTAPS report made the "unthinkable" even more unthinkable. It also sparked a debate among thinking scientists that continues to this day. Further studies confirmed the original work in broad outline, but there are many untested assumptions in the TTAPS study, including such things as how much soot is produced by urban fires, the size of soot particles, and how much soot would be injected into the upper atmosphere by smoke plumes. It was the purpose of the macabre California forest-fire experiment to refine those assumptions.

One hardly knows whether to be encouraged or depressed by such things as the California smoke experiment. On the one hand, one assumes that scientific knowledge of the unthinkable suffering that would ensue from the use of nuclear weapons would induce nations to think twice about the continuing buildup of weapons stockpiles, and certainly to refrain from using them. On the other hand, there is the terrifying lesson of history that what becomes thinkable usually happens.

CHAPTER 10

Seven Williams, Five Alexanders, Six Johns, Ten Vladimirs

There is something seductively attractive about a weapons system that would make us invulnerable to enemy attack. Unfortunately, the Star Wars space-based laser system proposed by ex-President Reagan and urged forward by many scientists has a built-in contradiction. To insure reliable protection against an all-out enemy attack the system must be incredibly complex, with mutually supporting and interacting parts. But the complexity of the system insures its occasional fallibility. A good analogy to SDI might be the Cell Wars defense system of the human body: The body's biological defense apparatus is amazingly successful, but it is not foolproof.

The human body is a society of cells. The integrity of the society is under permanent attack by viruses and microbes capable of causing disease or death. The first line of defense is the outer walls and moats: the skin, with its impregnable barrier of keratin, and the mucus membranes. Other exterior membranes of the body are flushed with fluids: saliva, tears, and nasal secretions, and, supplementing these watery defenses, the skin harbors a huge population of bacteria that do battle for the body the way pacified tribes on the marches fought for the Roman Empire.

Despite these surface defenses, hoards of viruses and microbes successfully penetrate the body. Sometimes they overwhelm the outer defenses by force of sheer numbers. Or they slip in quietly by an unguarded gate. They are masters of deceit and disguise. Once the enemy is within, sophisticated internal defense systems swing into action. The presence of an alien microorganism in the blood stream triggers chemical alarms that cause white blood cells to move to the site of the intrusion, where they attempt to engulf the enemy the way an amoeba engulfs its prey. If the foe is a virus, the infected cells release small proteins called interferon, like cries of warning. Interferon rouses surrounding cells and stimulates resistance to infection by the virus. Most effective of all the body's defenses are the lymphocytes, agents of the immune response. Lymphocytes are small, round, nondividing cells that are always on the alert. At any time there are as many as two trillion lymphocytes patrolling the human body. The huge number is crucial: Lymphocytes are very specific about what they recognize as an intruder. Each lymphocyte is prepared by evolution to respond to a particular alien. Recognition of a foreign agent causes lymphocytes to become active and start dividing. Offspring cells produce antibodies, which immediately go to work attacking the invader.

We live in a sea of alien viruses and microorganisms; they rain down upon us like a mass attack of enemy missiles. Many are harmless. Some are deadly. The body is protected by a stupendous array of traps, triggers, walls, moats, and chemical alarms. Various cells of the body act as patrols, sentries, infantry, and artillery to defend the integrity of the larger society. The Cell Wars defense system never rests. It is a thing of incredible complexity, vastly more intricate than SDI. And it works, without our awareness—unless and until something goes wrong. Cancer, AIDS, flu, and the common cold all represent failure, lapses of hardware or software, the "invincible" system run amuck, overwhelmed, corrupted from inside, blasted, bamboozled. The body, for all its vast array of defenses, is vincible.

And now, having used military images to describe the body's defenses, I am a bit abashed. It seems I have painted a picture of nature "red in tooth and claw," red with the vast senseless slaughter of Shiloh and Hiroshima, red with Yanomamo violence, cell set against cell, societies of cells locked in mortal combat, as if the antagonisms of nations were no more than the working out of some law of conflict that is intrinsic to life itself. And, yes, it is true. Not only tooth and claw, but also feather, eyelash, blush, and blossom are devices honed by evolution for trial by competition; each of these "inventions" gave their possessor an edge in the struggle to survive. But we should not overemphasize the Darwinian "struggle to survive" and "survival of the fittest." If there is a lesson to be learned from the complex, interacting defense system of the human body, it is that life is characterized mainly by cooperation. Our bodies are vast societies of mutually supporting cells, co-ops of highly evolved bacteria. The great thrust of evolution has always been toward "getting along." The multi-compartmented modern cell (with nucleus, chloroplasts, mitochondria, and flagella) is a symbiotic collaboration of simpler, more primitive, specialized cells. A gnat is a prodigious pulling together. A sponge is a confederation. The human body could not exist for even a minute unless the common good of a trillion cells took precedence over individual concerns.

The planet Earth may be the ultimate example of "getting along." Ecologist James Lovelock and biologist Lynn Margulis are exponents of the so-called Gaia hypothesis (Gaia was the Greek goddess of the earth). The hypothesis suggests that the planet, with its oceans, atmosphere, and biosphere, is a self-regulating organism prepared by four billion years of evolution to function harmoniously, of which human life is but one (perhaps disruptive) component. Gaia's integrity is a goal of antiwar activists and environmentalists who seek to ameliorate the effects of human violence against other humans and against nature. But it is a mistake, I think, to try and find in evolution the moral basis for nonviolence. Nature is neither moral or immoral; it is as often red as green. *Australopithecus*, Neanderthals, and the great ice-age mammals (to say nothing of trilobites, dinosaurs, and other vanished tribes) were victims of the amoral dynamic of evolution. Morality is not built into nature; it emerged at a certain level of cerebral evolution in the same way as photosynthesis, respiration, and sexual reproduction emerged at certain levels of cellular organization. Moral consciousness needs no justification; it exists.

For a year I had on my desk a clear glass sphere about three inches in diameter on a plastic stand. The sphere was two-thirds filled with water. The remaining volume contained air. A snip of sea grass, an alga, floated in the water, and four tiny pink shrimp swam lazily about. The sphere was completely sealed. With the exception of heat and light, this little society transacted no commerce with the outside environment. My glass-enclosed, shrimp-algae ecosystem was based on NASA-supported research, made available to the public under a technology transfer by Ecosphere Associates of Tucson, Arizona. Ecospheres are miniature worlds in which plants and animals live in balance with each other and all material resources and waste products are recycled.

The contents of the Ecosphere are not as simple as they appear. The sea grass and the shrimp are obvious flora and fauna, but these two species alone are not sufficient to sustain life in a sealed container. There are perhaps as many as a hundred species of life in the glass globe, mostly microorganisms invisible to the eye. The plants make oxygen and food for the animals from sunlight, carbon dioxide, and inorganic chemicals. The animals breathe in oxygen and expel carbon dioxide. They eat the plants and produce organic wastes. Bacteria oxidize the organic wastes and produce more carbon dioxide and inorganic chemicals. The precise mix of organisms is crucial for the longterm success and beauty of the system. For example, ammonia is a normal waste product of shrimp, but becomes poisonous in high concentrations. At least two kinds of bacteria are necessary to convert the ammonia into useful nitrite, thus keeping the nutrient and energy cycles going. Shrimp have stayed alive in closed shrimp-algae environments for as long as seven years, and, so far, this is a record for any of the higher forms of life. If humans are ever to establish permanent colonies on the moon or Mars, closed-ecosystem research will show the way. The cost of transporting enough food, water, and oxygen from the earth to sustain a space colony would be prohibitive. It will be necessary to provide the colony with the ingredients necessary for a self-sustaining, recyclable environment, similar to the one in my glass sphere, and this is why NASA invests time and money in closed-ecosystem research.

The earth's biosphere is roughly a billion trillion times more voluminous than the little three-inch sphere that rested for so long on my desk. Research has shown that there is a direct relationship between the size of a closed ecosystem and its ability to sustain a balance of life, and that therefore the earth's biosphere is vastly more stable than the Ecosphere. Nevertheless, it would be a mistake to imagine that the biosphere of the earth is so large as to be invulnerable to disruption. Episodes of mass extinctions in the fossil record (trilobites, dinosaurs) provide ample evidence that the equation can sometimes go wildly out of balance. The clear-cutting of tropical forests, the unrestrained burning of fossil fuels, and (most dramatically) nuclear war are examples of human activities that can significantly alter the equation of matter and energy that has sustained life on earth for nearly four billion years.

Now I will reveal the fate of my Ecosphere. I went away on a Christmas holiday and closed the door to my office. I had forgotten that I had turned off the heat to the room. When I returned, the four tiny pink shrimp had become two, and the two remaining shrimp showed few signs of life. The extent of the tragedy for the dozens of *unseen species* could only be guessed. With the return of warmth, the two remaining shrimp slowly began to revive, but never quite regained their old vigor. And then came catastrophe from on high, the equivalent, I suppose, of the impact with earth of a massive asteroid from space, or nuclear war—disaster sudden and unequivocal. I accidentally knocked the sphere off the desk. A smash, a splash. Apocalypse. Extinction.

I did not buy another Ecosphere. Perhaps I no longer trusted my competence to play God, even for a world I could hold in the palm of my hand. But my experience with the Ecosphere deepened my sense of responsibility for the larger closed ecosystem which is Earth. Human intelligence and technological prowess are new factors in the equation of evolution. The speed of our interventions into the environment far outstrips nature's capacity to respond constructively, depending as it does upon the slower calculus of genetic transmutations. My four pink shrimp depended for their existence upon a sustained balance of interdependent organisms. When that balance was disrupted, admittedly from without, their little world came to a calamitous end. Their fate could be our own.

Early in 1988, a friend gave me a new poster from the Smithsonian Air & Space Museum called "Space Explorers," a compilation of small portraits of all persons who had spent at least one earth-orbit in space. I was surprised. I knew that the number of people who had gone into space was large, but seeing them massed together impressed upon me the special nature of this remarkable confraternity.

At the time the poster was published, 204 human beings had experienced the earth from the overview of space. Seventy-six of those persons were launched into space by a Soviet vehicle. The remainder were passengers on an American craft. Soviet cosmonauts and American astronauts are designated on the poster by a red or blue name bar under each portrait. If it were not for the colors you wouldn't know East from West. The faces are strikingly similar: bright, eager, confident, and, to put it simply, human. The youngest space explorers, a Soviet man and woman, were twenty-six at the time of their flights. The oldest, an American, was fifty-eight. Ten women had orbited the earth by early 1988. Among the faces on the poster are representatives of many races. Nationalities include a Bulgarian, a Canadian, a Cuban, a Czech, a Dutchman, two Frenchmen, an East German, three West Germans, a Hungarian, an Indian, a Mexican, a Mongolian, a Pole, a Romanian, a Saudi Arabian, a Syrian, and a Vietnamese. Some of the names have the quality of legend, especially those from the early years. Yuri Gagarin. John Glenn. Valentina Tereshkova. Neil Armstrong. Others are only vaguely familiar. There are seven Williams. Five Alexanders. Six Johns. Ten Vladimirs. Just ordinary folks with the right stuff.

The portraits are arranged chronologically, according to the time of the space flight, and patterns of color emerge on the poster, blocks of red or blue. At the top, the early sixties are dominated by Soviet red. Then a band of American blue, lofted into orbit by Gemini. Red again with the Soyuz spacecraft as Americans prepared Apollo. More blue with the grand Apollo adventure to the moon, and with Skylab. The middle of the chart is packed with red, steady Soyuz, the reliable Soviet workhorse. Near the bottom of the poster, a massive bank of blue, the Shuttle, seats for eight, loaded like a bus. Then, red again, in the aftermath of the *Challenger* tragedy, Soyuz perfected, the Soviet turtle for the American hare in the race for the stars. But the faces on the poster do not evoke a sense of deadly competition. The smiles are too open, the eyes too uniformly full of curiosity. What comes across is the potential of space exploration for bringing the world together, across the boundaries that divide us.

Science writer Frank White has outlined how the view from

space can contribute to positive human evolution in a book called *The Overview Effect*. He lets the space explorers speak for themselves. Here is Yuri Gagarin, the first human to enter space: "Trembling with excitement I watched a world so new and unknown to me, trying to see and remember everything." And Prince Sultan Salmon al-Saud of Saudi Arabia, who flew on the shuttle: "I think the minute I saw the view for the first time was really one of the most memorable moments of my entire life. I just said in Arabic, 'Oh, God,' or something like 'God is great.' It's beyond description." The "overview effect" was similar for most of the space explorers. The response of Michael Collins, who orbited the moon while Armstrong and Aldrin took the first steps in moon dust, is typical: "I think the view from 100,000 miles could be invaluable in getting people together to work out joint solutions, by causing them to realize that the planet we share unites us in a way far more basic and far more important than differences in skin color or religion or economic system. The pity of it is that so far the view from 100,000 miles has been the exclusive property of a handful of test pilots, rather than the world leaders who need this new perspective, or the poets who might communicate it to them."

More than two hundred human beings have experienced the unique, transforming view of the planet Earth from space, visually encompassed the planet in the same way as I once held the three-inch Ecosphere in my hand: blue-white Gaia suspended in the black of space, sheathed in her glistening skin of water and air. The rest of us can be admiring, or envious—or we can learn from their experience. Imagine if you will a Soviet-American collaboration on the next great human adventure in space, the exploration of Mars early in the next century. An international crew of men and women fluent in each other's languages. Hardware built partly in this country and partly in the Soviet Union, harnessing competition in the interests of quality and safety. Star ships replacing Star Wars. Orbiting launch stations instead of spy satellites. Exploring a new planet, rather than destroying this one. A further step in evolution's thrust toward cooperation and getting along, the common good taking precedence over individual concerns. The faces on the Smithsonian poster display an openness and a readiness to embrace such a venture, if only we can find the leaders with the imagination to lead the way.

CHAPTER 11

The Cicada's Song

When I was a boy growing up in Tennessee I "borrowed" my uncle's .22 rifle and went hunting with friends. My very first shot brought a gray squirrel tumbling down through the branches of a tall pin oak. The squirrel lay on the ground at my feet, convulsed with pain, its belly pierced by a neat red hole. I watched, paralyzed by the enormity of what I had done, until one of my friends dispatched the animal with a single crushing blow of his rifle butt. The sight of the suffering squirrel, and then of its broken cadaver, moved me deeply. Never since that day have I gratuitously injured another living thing. I tell this story to indicate the kind of deep emotions that can color our thinking about the highly charged issue of animal rights and scientific experimentation with animals.

The animal-rights movement and current practices of scientific research are on a collision course. Opposition to the use of animals in research is not new. What is new is a growing public interest in the philosophical issue of animal rights, which draws life from the ecology movement, and the ability of antivivisectionists to attract powerful political support. In America, the issue was pressed forcibly upon the public consciousness in 1984 when animal-rights activists broke into the head-injury research lab at the University of Pennsylvania and stole videotapes of violent, cranium-bashing experiments with baboons. Copies of the tapes were widely disseminated and evoked a furor of outrage. This particular use of animals clearly violated the public sense of what is morally acceptable. The University of Pennsylvania was subsequently fined four thousand dollars for abuse of the animals, and the episode led to a tightening of government and professional guidelines for animal research. In 1985 Congress enacted new laws aimed at insuring the welfare of laboratory animals, but the issue did not go away. Inspired by animal-rights philosophers such as Peter Singer and Tom Regan, activists continue to press for more restrictive legislation on experimentation with animals.

In 1985 more than one hundred million animals were killed worldwide in the course of scientific research—mostly rats, mice, and guinea pigs, but also large numbers of cats, dogs, and monkeys. In this country the figure was something like twenty million. To put these numbers into perspective, Americans kill about four billion animals each year to fill their stomachs, and the number of animals used in research, particularly vertebrates, continues to decline. As an informed citizen, what should be *my* attitude about all of this? I would like to think of myself as compassionate toward my fellow creatures. But I do eat my share of them. I have no hesitation about swatting health-threatening insects. I am as outraged as the next person when laboratory animals are treated cruelly or callously, but I am grateful for the advances in public health and safety that have accrued from the responsible use of animals in education and research.

There are no simple answers to this complex moral question, but scientists should be among the first to nurture public awareness regarding the use and misuse of animals in research and education. No one better understands the many and subtle ways we are bound to one another—bacteria, fruit flies, white rats, baboons, pink shrimp, and clumsy humans—than those who study the life sciences. Knowledge is a prerequisite for love, and those who best understand the interdependence of life should be exemplars of compassion. Compassion, like life itself, is a seamless web. Thoreau wrote this about animals and their rights: "I have found repeatedly, of late years, that I cannot fish without falling a little in self-respect . . . I have skill at it, and, like many of my fellows, a certain instinct for it . . . But always when I have done I feel that it would have been better if I had not fished. I think that I do not mistake. It is a faint intimation, yet so are the first streaks of morning."

Herewith, an animal-rights story in several parts:

CHAPTER 1: Jenifer Graham, a fifteen-year-old high school student in Victorville, California, refuses to dissect a frog in biology class. Jenifer is a vegetarian and opposed to the unnecessary use of animals for food or research. She asks for permission to learn frog anatomy from a model or a computer simulation. School officials say she must cut up the frog or get out of class.

CHAPTER 2: After a lively controversy, Jenifer gets a lowered grade, and her refusal to dissect the frog is noted on her transcript. With the help of animal-rights activists, she sues, claiming violation of her constitutional rights.

CHAPTER 3: Enter Apple Computer, which markets a computer program called "Operation Frog." Jenifer appears in an Apple television commercial saying: "Last year in my biology class, I refused to dissect a frog. I didn't want to hurt a living thing. I said I would be happy to do it on an Apple computer. That way, I can learn and the frog lives. But that got me into a lot of trouble, and I got a lower grade. So this year, I'm using my Apple II to study something entirely new—constitutional law."

CHAPTER 4: Scientists and educators react with dismay to the Apple commercial. The California Biomedical Research Association urges its members to write Apple president John Scully protesting the commercial. It is claimed that the message is "offensive" to scientific educators, and "advances the cause of fanatics." Apple pulls the commercial.

And that is perhaps where the story should end. Animal rights is an inflammatory issue that often seems to be fueled more by emotion than by reason—on both sides. But into this thorny thicket steps Daniel Koshland, editor of *Science*, the weekly journal of the American Association for the Advancement of Science.

CHAPTER 5: In an editorial, Koshland attempts to reduce the positions of animal-rights activists to absurdity. "There are a number of clues about the insides of a frog," he writes, "such as that it arises from a tadpole, that it causes warts, and that it may turn into a Prince Charming when kissed by a beautiful princess. From such data, a moderately well-trained student should easily be able to deduce what the interior of a frog looks like"—without dissection. Cutting up a real frog, says Koshland, would reveal a stomach full of flies, mosquitoes, and small grasshoppers, thereby subjecting students to the harsh reality of a world where animals eat animals. To spare students this trauma, the editor of *Science* suggests a computer simulation that fills the frog's stomach with more consoling items—such as potato chips and soda pop. And why not, he continues, let the Food and Drug Administration replace costly and time-consuming animal testing of drugs with computer programs. Why not replace mousetraps and fly swatters with computer programs, or at least enact legislation requiring that flies be anesthetized before they are swatted. Carried to its logical conclusion, Koshland implies, Jenifer Graham's position should preclude even the eating of plants, which after all are living things. The obvious answer, he suggests, is to genetically engineer human beings that can photosynthesize their own food, and dubs them *Homo photosyntheticus*.

In fairness, I should stress that Koshland's editorial was directed to scientists, not the general public. And the issue of animal rights can certainly do with a little humor. Still, I have the feeling that the derisive tone of Koshland's piece did little to advance the case for animal experimentation in science. And I also have a sneaking admiration for Jenifer Graham. Here at least is a high school student who abhors what she perceives to be cruelty to animals, and who is willing to stand up for her principles. She is certainly not the only student who has found the dissection of frogs and cats morally repugnant or emotionally distressing. Nor am I unsympathetic to Apple Computer's attempt to sell a program that simulates frog anatomy. Clearly, animal dissection is an essential part of the education of physicians, veterinarians, and research scientists in the life sciences, but it is not equally obvious that dissection is an indispensable part of a high school biology course, or even college Biology 101. A pedagogically sophisticated computer simulation might conceivably teach more about frog anatomy than a real frog, and even about the contents of a frog's stomach. Whatever educational advantages accrue from the dissection of animals must be balanced against the need to instill in students a reverence for life on a planet where all creatures live in interdependence, and where one creature bears a particular burden of responsibility. The balance is not as easily found as Koshland's editorial suggests.

The animal-rights movement continues to gain adherents, and scientists might feel that to bend on an issue like Jenifer Graham's frog puts more essential uses of laboratory animals at risk. Maybe so. But it seems to me the best defense against those who would deprive science of the opportunity for animal experimentation is a self-imposed policy that seeks to rigorously minimize the unnecessary use—and, certainly, the misuse—of laboratory animals, and a willingness to listen to those who suggest reasonable alternatives to dissection or vivisection.

Not long after I killed the squirrel with my uncle's rifle, I discovered among the books in my family's library J. Henri Fabre's *Insect Adventures,* stories translated from the voluminous works of the great nineteenth-century French entomologist retold for young readers. Not much retelling was necessary. All of Fabre's books are delightfully and simply written, and popular with readers of all ages. They remain, after a century, the best and most engaging introduction to the world of insects. I am certainly not the only person who was nudged toward the study of science by one of Fabre's books.

Fabre made bugs—ordinary bugs of the household and garden—as exciting as the great beasts of the African veld. He told stories of their nestings and matings, their languages and societies, and their roles as predators and prey, all based upon his own careful observations. But in spite of popular success, Fabre was never made a welcome member of the scientific community. His folksy, literary prose style was resented by fellow entomologists. They were further put off by his resistance to dissection and laboratory experiments. Stymied in his career, Fabre never advanced beyond an assistant professorship at a tiny salary. Fabre believed that the methods of science must be consistent with our motives for knowing, and for this reason alone he resisted laboratory experimentation with animals. His method was to enter as intimately as possible into the lives of the creatures he studied. His laboratory was the field. "I make my observations under the blue sky," he wrote, "to the song of the Cicada."

As counterpoint to Fabre's method, consider two articles published in the journal *Science*. The articles are interesting both for what they report about insects and as instructive examples of the establishment's way of doing animal research. Both reports describe flies that have evolved physical features and behaviors that let them mimic their chief predators, jumping spiders. The flies' wings are banded with stripes that resemble the legs of the spiders. When threatened, the flies wiggle their wings so as to mimic the spider's movement and gait. Presumably, the point of this little ruse is to avoid being eaten. Mimicry is not unusual in the world of insects. Insects have evolved with body shapes and colors that mimic flowers, twigs, leaves, pebbles, other insects, and even bird droppings, the better to escape detection by predators. The sort of mimicry described in the two *Science* articles, where the prey mimics the predator (a "sheep in wolf's clothing"), has been only rarely reported.

In the first of the two reports, researchers at Simon Fraser University in British Columbia describe experiments with snowberry flies. The predator, a jumping spider, was admitted into a Plexiglas "arena." Within the arena, a snowberry fly was confined under a clear glass dome and the spider's response was observed. As expected, the spider responded to the mimic fly as if it were another spider. The experiments were repeated with jumping spiders under the dome, with ordinary house flies, and with snowberry flies whose wing stripes were obliterated with a marking pen. The spiders acknowledged the other spiders in a spiderly fashion, but behaved aggressively toward house flies and the snowberry flies that had been deprived of their disguise. Clearly, coloration mimicry, and not some subtle chemical or behavioral signal, determined the spiders' responses.

The second report, by three American researchers, describes experiments with another species of fly (*Zonosemata*) with leg-like markings on its wings and false eyespots on its abdomen that mimic a jumping spider. Five types of "prey" were introduced into the company of hungry spiders: (1) Normal *Zonosemata;* (2) *Zonosemata* whose leg-banded wings had been cut off and replaced (using Elmer's glue) with the wings of ordinary houseflies; (3) *Zonosemata* whose wings were replaced with the wings of other *Zonosemata;* (4) ordinary houseflies; and (5) houseflies equipped with glued-on *Zonosemata* wings. The spiders acknowledged *Zonosemata*—with their own wings or the wings of another *Zonosemata*—in friendly fashion, and behaved aggressively toward all other categories of prey. Both leg stripes and distinctive wing movements were necessary to fool the spiders. Predators other than spiders attacked *Zonosemata* as readily as any other prey. Apparently, *Zonosemata*'s distinctive coloration evolved as a specific deterrent to attack by jumping spiders.

J. Henri Fabre would have been fascinated to hear of these wonderful mimic flies, furiously waving their false legs, flexing their eye spots, and trying for dear life to look like jumping spiders. He would certainly have considered the tiny deceptions of the mimic flies one more remarkable insect adventure, to be patiently observed and exhaustively described, even if never fully understood. But I suspect Fabre would have been less enchanted with the apparatus of the experiments—the Plexiglas arena, marking pens, scalpels, and glue—and with the snipping and marking of wings. His motive for knowledge was love and reverence for life. He wore himself out discovering the secrets of insects, on his knees in the grass under a clear blue sky, ears ever alert to the cicada's song.

CHAPTER 12

One Thrips, Two Thrips

I sing the praises of pear thrips, flip-flop birthing insects, scourge of the syrup makers. I sing the praises of sea-daisies, pinhead sized medusas, sole members of biology's newest class. I sing the praises of *Escherichia coli,* industrious and versatile bacterium, footless foot soldier in the cause of science, commensal friend. But before I single out these three from among the planet's ten million species of life, let me pause to acknowledge the quite remarkable fact that we are the only one of those ten million creatures that is curious about all the rest. Moreover, we are curious about the very machinery of life itself, the unseen molecular dynamics that energize the heart of every cell, the lock-and-key appliances of proteins, and twisted-helix concatenations of chemical units contrived by evolution that let, for example, *Zonosemata* and snowberry flies mimic their mortal enemies. For this uniquely human agenda of praise and understanding, it is not enough to rummage in the grass, like Fabre, observing with a poet's eye the collective antics and domestic arrangements of the beasts. We are required by the unexamined protocols of science to test and tinker, sometimes dissect and destroy. Our motive for knowing is unrestrained curiosity, and our method seems to be whatever works. We have (some of us) ceased to be the quaintly vertical, mammoth-bashing *Homo erectus:* self-centered, single-minded, capable of obliterating entire species of animals in pursuit of a piece of fur or a chunk of meat. We have become instead, by a brain-fulling quirk of genes, *Homo curiosus:* hesitant, awestruck, uncertain about our proper relationship with the other inhabitants of Earth, desperately desiring to understand what it all means.

A newspaper story: Vermont maple syrup producers are locked in a life-and-death battle with the pear "thrip," a tiny but voracious insect that defoliates maple trees, which is not good for the sap.

First, let's get one thing straight. There's no such thing as a thrip. The singular of thrips is thrips. The plural of thrips is thrips. As Dr. Seuss might say, "One thrips. Two thrips. Red thrips. Blue thrips." But then again, the thrips that afflicts maples is neither red nor blue, but blackish or whitish depending on the stage of its life cycle. Dark adult thrips spend the winter underground, emerging in early spring to feed on buds, blossoms, leaves, and young fruits of various trees, including pears and sugar maples. They deposit eggs in leafstalks. Pale young thrips emerge from eggs, eat, get fat, then fall to the ground, where they take up residence in the soil and lie dormant for the summer. Pupation takes place in late fall, whereupon thrips go back to sleep for the duration of the winter.

It's a simple life, mostly spent snoozing, for an almost invisibly small insect with a Dr. Seuss sort of name. But this essay is not about thrips; it's about humans, about curiosity, and about the Dr. Seuss in all of us. One of my favorite pastimes, in season, is to plunk a wildflower onto the stage of a dissecting microscope and go exploring. No Darwin in the Galápagos or Humboldt on the Amazon ever surveyed a richer fauna than that which exists microscopically on almost any plant. A dandelion can be a tropical forest for the naturalist armed with even a modest degree of optical magnification. Within the stamens and corollas of every meadow weed lurks—as J. Henri Fabre knew—an astonishing zoo of beasts. Now I wouldn't know a pear thrips from a wheat thrips or any other kind of thrips (there are hundreds), but I know a thrips when I see one. Adult thrips have four fringed wings, like tiny oars with eyelashes, and often I've seen one skulking about in a flower head. Magnified ten times, even the wimpishly named thrips presents a formidable aspect. Eye to eye with one of these monsters you begin to understand how a thriving throng of thrips could trash a maple.

No sooner had I read the newspaper story on thrips in Vermont than I came across another thrips story in the journal *Nature*. "Facultative viviparity in a thrips," it was called, by Bernard Crespi of the University of Michigan. Crespi discovered a species of thrips in which females lay eggs (oviparity) or give live birth (viviparity) or both. The oviparous offspring are female, and the viviparous offspring are male. During a given bout of reproduction it's all one way or the other, but long-lived females can change modes and somehow the ratio of sexes balances out. Crespi's switch-hitting female thrips is apparently the only known animal capable of "choosing" her mode of reproduction in response to subtle signals from within or without.

This sort of eccentricity is the delight of evolutionists. How and why did such bizarre behavior evolve? What is the adaptive value of being able to switch reproductive modes? Why is being born live good for males but not for females? Crespi posits plausible explanations, using the language of Darwinian logic, but I wonder if an explanation by natural selection is really necessary. Thrips run the gamut of reproductive strategies. Some species lay eggs, some give birth to live young, and one, at least—Crespi's thrips—has it both ways. Nature seems to have a niche amongst its more than ten million species for almost everything. Not even the wildest product of Dr. Seuss's imagination—the Moth-

Watching Sneth, or the Grickily Gractus (that lays eggs on a cactus)—is stranger than creatures that actually exist. The egg-laying/live-birthing Flip-Flop Thrips is a case in point. So is the sweet-toothed Syrup-Sipping Thrips of Vermont.

And now I'm warming to my subject. The naturalist Donald Culross Peattie wrote: "The world, from a weevily point of view, seems to exist for the weevils." He might as well have said that sugar maples, from the thrips' point of view, exist for thrips. And, of course, he's right: it's all a matter of point of view. I'll draw my moral like this: From the human point of view, thrips exist for humans. Why does a biologist like Bernard Crespi spend all that time with nylon mesh bags filled with dead oak leaves and live thrips, counting offspring, sexing baby thrips, and determining reproductive mode? Some thrips are agricultural pests, and one can argue—especially when applying for research grants—that studying the reproductive behaviors of thrips will have an economic payoff. But I doubt if such practical concerns have much to do with Crespi's motivation, or that of any other entomologist. He's doing in a serious way what I do more casually with my microscope—following his curiosity into the bizarre Dr. Seussian world of biological diversity.

There are more than ten million species of life on this planet, and we are the only one curious about all the rest. If somewhere there's a bird that watches moths or lays eggs on a cactus, we want to know about it. If there exists a creature of bivalent reproductive mode, we want to know that, too. Not even the lowliest of Earth's creatures—not even the fringe-winged thrips—escapes our interest.

Another report from the British journal *Nature:* "A new class of Echinodermata from New Zealand." The authors describe an animal previously unknown to science, nine of which were discovered on waterlogged wood dredged up from the ocean off the New Zealand coast.

Echinodermata are the phylum of spiny marine animals that includes starfish, sand dollars, and sea urchins. No new class of living echinoderms has been found since 1821. The new animal from New Zealand differs from all other living echinoderms in having a double rather than single ring of canals in its water vascular system. The tiny creature (it is not much bigger than this letter *O*) has been informally dubbed a "sea-daisy." The official designation of the new class is Concentricycloidea. The genus and species are *Xyloplax medusiformis*. In science, living creatures are classified according to a system that looks like a family tree. As one moves inward along a branch of the tree from twig to trunk—species, genus, family, order, class, phylum—the likelihood of finding a new member grows increasingly remote, so the New Zealand discovery is no small thing.

The report in *Nature* follows the standard format for announcing such discoveries. First are given the names for class, order, family, genus, and species, with diagnosis and etymology (but not the evocative nickname "sea-daisy"). Next comes a precise accounting of the circumstances of discovery. Finally, the authors present a detailed anatomical description of the new animal. Missing from all such reports are the thoughts and emotions of the discoverers. Of them, we know only their names and their place of employment. By self-imposed prohibition, scientific reporting allows little room for human emotion. We must rely on our own imaginations to supply the feelings that were present at that magical moment when the two New Zealanders and the Australian recognized that they had found something significantly new: elation, laughter, amazement, relief, joy in the successful hunt, surprise, gratitude, the electrifying tingle in the spine.

The novelist Vladimir Nabokov, who was also an accomplished lepidopterist, described the pleasure of discovering a new butterfly as being like that of a child who has learned to ride a bicycle.

The dream at the back of every lepidopterist's mind, he wrote, whether climbing a mountain in New Guinea or crossing a bog in Maine, is of capturing the first specimen of a species unknown to science. Nabokov had the good fortune and intense pleasure of discovering several new species and subspecies of butterflies. Butterflies belong to the phylum Arthropoda, a category that includes the insects, spiders, and crabs. Half a million species of insects alone have been described, and some zoologists believe there are ten million more waiting to be discovered, some, no doubt, with unusual habits, such as the oviparous/viviparous thrips. Echinoderms are a more restricted phylum; only some six thousand living species are known. To discover an echinoderm so different as to require the creation of a new class—Concentricycloidea, the sea-daisy—must have been a very great thrill indeed, even if the emotion is absent from the scientific report.

Literary reports that convey the human aspect of discovery, such as Nabokov's, are rare. I'll recount one more, from Edmund Gosse's famous autobiography *Father and Son*. Gosse's father was a zoologist and a writer of books on natural history. On June 29, 1859, before Edmund Gosse was ten, he accompanied his father on a collecting expedition along the Devon shore. He found a creature with which he was unacquainted, and ran with overbrimming pleasure to announce the discovery to his father. It turned out to be not only a new species but a new genus to be added to the British fauna, a sea anemone now known as *Phellia murocincta*, or walled corklet. Gosse's recollections of "those delicious agitations by the edge of the salt sea wave" vividly suggest the excitement that comes when a new twig has been added to the tree of life.

It is not hard to guess that Alan Baker, Helen Clark, and Frank Rowe, the three echinoderm specialists from Down Under, looked upon their nine little daisylike specimens with something akin to delicious agitation. J. Henri Fabre gave a poet's expression to those delicious agitations, and for his trouble was spurned by the scientific establishment. The prosaic, passionless formulas of scientific reporting serve an important purpose: They maintain science as a communal enterprise, free of the prejudices of nationality, race, religion, politics, and personality that have plagued so many human enterprises. It is the ideal of science that facts and theories stand unembellished by personality, to be judged independently of the quirks and agitations of their discovery or creation. As we read in *Nature* the matter-of-fact account of the discovery of *Xyloplax medusiformis* (or sea-daisy), we must supply for ourselves the missing human dimension, the heady kick of discovery, the rush of adrenalin, the once-in-a-lifetime thrill that Nabokov called the "ardent and arduous quest ending in the silky triangle of a folded butterfly lying on the palm of one's hand."

Of the more than ten million species of life on this planet, which do we know best? Ourselves? *Homo curiosus*? Not at all; the human frame remains wrapped in mystery; our big brain and complex nervous system, especially, render us the Dark Continent of biology. What about the famous laboratory white mouse, the subject of countless biological and behavioral experiments? The white mouse is well understood, yes; but not nearly so exhaustively described as *Drosophila*, the fruit fly. For almost a century, *Drosophila* has been bred in laboratories by the billions for research on genetics, sensory mechanisms, neural networks, biological rhythms, learning and memory, and various behaviors. There is hardly a biology or medical lab in the world that does not have racks of bottles buzzing with the amatory songs of this rapidly reproducing, prolific ally in the quest for the secrets of life. But *Drosophila* does not take the prize as the subject of choice in scientific research. That honor goes to *Escherichia coli*, bean-shaped bacterium, multitudinous inhabitant of the human intestine, best understood (as I shall now reveal) of all God's creatures.

Bacteria are the most primitive forms of life on earth. *E. coli* is about as simple as a creature can be and still be alive—a microscopic pinch of protoplasm in a membrane. It has a single chromosome and a loop of DNA about as big (if stretched into a circle) as the dot on this letter i. (By contrast, the DNA in a human cell is a thousand times as long.) *E. coli* moves. It feeds. It reproduces. It recognizes and communicates with its own kind. In a primitive sense, it can even remember. It gets from here to there by wriggling whiplike appendages; or screwing them, actually, like propellers. It has, such as it is, a life of its own. It is the least of creatures. It is ubiquitous.

E. coli is microscopic. A million of them laid end to end would make a line only as long as my arm, but there are enough *E. coli* inside of me now (citizens of my intestines) to form a line that would stretch from Boston to San Francisco. A significant part of my body weight is not me at all, but my burgeoning population of bacteria. As house guests go, my *E. coli* are not unwelcome. They produce, I am told, certain useful vitamins. They sometimes devour other disease-causing microorganisms. But mostly they just go about their business, happily sharing my space, doing me very little good or harm. The technical term for our relationship is *commensal:* literally, "eating at the same table."

E. coli is far and away the most intensively studied of any living organism. I checked a recent volume of *Biological Abstracts,* a journal that indexes biological research. There were five times more references to *E. coli* than to any other species. (*Drosophila* is a distant runner-up.) This little bacteria has no secrets. Its genes have been exhaustively mapped. Its proteins have been cataloged. Its invisibly small self has been poked and probed and scrutinized in a thousand ways. *E. coli*'s simplicity makes it attractive to science; the little bacterium is the distilled, infinitesimal essence of life, the versatile white mouse of microbial research. It is hardy. It multiplies itself every twenty minutes. Genetic researchers, especially, have found *E. coli* to be an ideal subject. It is possible to snip little bits of gene from other creatures and splice them into *E. coli*'s DNA, thus tricking the bacteria into doing things it would otherwise know nothing about (a technique called recombinant DNA research). *E. coli*'s genes have been thus engineered in ways to cause the creature to unwittingly manufacture insulin, interferon, and human hormones, immensely valuable substances that are otherwise extremely difficult to obtain. For example, the human brain hormone somatostatin is a substance that inhibits the secretion of the pituitary growth hormone. Researchers who first isolated somatostatin needed half a million sheep brains to collect five milligrams of the substance. A two-gallon culture of genetically altered *E. coli* can quickly produce the same amount. (Of course, there is a dark side to the restructuring of bacterial genes. In the early days of genetic research it was feared that a dangerous variant of reengineered *E. coli* might escape from the lab and cause disease among humans. Stringent measures were taken to insure the containment of the bacteria. Nowadays, most *E. coli* research employs a deliberately crippled strain of the bacteria that cannot live except in the specific conditions of the lab.) With the success of these genetic engineering experiments, a new industry was born. It will not be long before factories employ vast numbers of genetically altered *E. coli* laboring in blind response to their tinkered-with genes to produce useful products for humankind: medicines, insecticides, fertilizers, and—if we are not watchful—weapons of war.

I cannot help but feel a bacterium-sized twinge of conscience about the cavalier way we manipulate *E. coli*'s genes. *E. coli* serves, in laboratory vials and teeming factory vats, because *Homo curiosus* wants to know. It is inevitable that scientists will stick their fingers into the web of life, for the need to know is as compelling an appetite as the need for food, clothing, or shelter. But as we poke and probe the silky net of life we should stay attuned to the resiliency of the web, which has, after all, no top or bottom, no center or perimeter, only a plenitude of necessary strands.

CHAPTER 13

The Bluebird of Happiness

In autumn, Thoreau listened for the sound of the loon on Walden Pond, "a wild sound, heard afar and suited to the wildest lake . . . a long-drawn unearthly howl, probably more like a wolf than any bird." We need, he said, "the tonic of wildness." The loon's voice is a heady distillation of the tonic of wildness, a rich intoxicating liqueur. But already, even in Thoreau's day, the loon's continued existence was threatened by hunters. The mere rumor of a loon on Walden Pond was enough to bring Thoreau's sportsmen neighbors flocking from the town, "in gigs, on foot, two by two, three by three, with patent rifles, patches, conical balls, spy-glass or open hole over the barrel." The woods near Thoreau's cabin reverberated with the crack and bang of domestic ordinance. Loons never fared well in proximity with humans. They have retreated with their wild music to northerly woodlands. You are unlikely to hear the cry of a loon on Walden Pond today.

Recently, a migrating loon stopped on one of our campus ponds—a fatal, unfortunate visit. A student playing soccer on an athletic field near the pond said she heard a "gosh-awful noise." She looked up in time to see a great bird smash into the bleachers with full force. The bird had lifted from the pond and was circling to gain speed and altitude. A loon is a powerful swimmer and diver, but it has short wings for a bird of its body weight and its body is not well balanced for life out of the water. A loon cannot take flight from land, and must struggle to lift itself from water. Our visitor was engaged in that struggle when it hit the bleachers. Of course, from the loon's point of view, the bleachers had no business being there; they were as dangerous an intrusion into the loon's wild sphere as the barking firearms of the good citizens of Thoreau's Concord.

I saw the loon in death. Its broken carcass was collected by a campus naturalist and placed on a tray in a laboratory refrigerator. The Department of Fisheries and Wildlife was notified. Those of us with an interest in such things made our melancholy pilgrimage to the lab; the experience, I suppose, was rather like going to the morgue to identify the body of a departed loved one. The loon was recognizably part of ourselves; something beautiful, natural, feral, and wild. Cold and leaden on a plastic tray.

Fowling arms and sports bleachers are not the only human threats to the loon's continued existence. Summer cottages and motorboats on our wild northern lakes disturb the peace that loons require to breed successfully. Many loon chicks are lost in the egg stage to the predations of gulls and raccoons, two animals that flourish in untidy human environments. No matter how deeply into the woods the loon flees, the effects of human activity follow him. Industrial pollution is carried by wind to the remotest waters of Canada and Maine, where it falls as acid rain. Acid rain leaches metals from the soil, which are then carried into lakes and ponds and are incorporated into the food chain (loons are particularly susceptible to mercury poisoning). But the consequence of acid rain can be even more direct. Says ornithologist William Scheller: "Acid rain kills fish. Loons eat fish. No fish, no loons . . . it's as simple as that."

Good housekeeping by lakeside residents is essential if the loon is to make a comeback. Conservationists have embarked upon a program of public education to tidy up the environment and help the loon find elbow room on northern lakes. Artificial islands have been constructed in some of the larger lakes to provide the bird with secure nesting grounds. Most important, as an endangered species the loon enjoys the protection of law. The Department of Fisheries and Wildlife expressed a keen interest in the bird that flew into our bleachers. Environmental police officers came quickly to the campus to collect the carcass. I felt something hugely sad as the bird was taken away. That one loon's death was emblematic of the loss of something greater, something wild and free. I remembered something Thoreau had written: "In wildness is the preservation of the world." Human progress has put the wilderness at risk. Forests fall. Grasslands are paved over. Ponds fill with acid rain. The web of life is pared back, diversity is abbreviated. Science, the Janus-faced god of knowing, is the instrument of both dismemberment and preservation.

In his journal, Thoreau tells how he disturbed a loon while rowing his boat on Walden Pond. The loon cried out, wrote Thoreau, "as if . . . calling on the god of loons to aid him." The gentle naturalist of Concord was no great threat to the loon. Acid rain and the callous exploitation of the environment pose more serious and immediate dangers. If there is a god of loons, he has his hands full today.

There is nothing quite so final as extinction. We will never see the likes of Neanderthals again. Woolly mammoths, saber-toothed tigers, and passenger pigeons are gone forever. But within this catalog of diminished diversity we are offered one token of grace. The ivory-billed woodpecker lives!

That is the announcement made in 1986 by ornithologist Lester Short of the American Museum of Natural History. Short and his colleagues sighted several ivory-billed woodpeckers in the mountain forests of Eastern Cuba. Most ornithologists had considered the bird to be extinct. The ivory-billed woodpecker was once common in Cuba and in the southeastern United States. It is a large bird, with a wing span of nearly three feet. It has shiny black plumage marked by distinctive white stripes, white-tipped wings, a magnificent scarlet crest, and of course the dagger-shaped ivory bill that gives the bird its name. In Audubon's painting of a male and two females, the birds have something of the look of a family of gaily colored pterodactyls. During his travels in the American south, Audubon frequently encountered ivory-billed woodpeckers. Like them, he loved the solitude of wild, forested places.

The ivory-billed woodpecker's range once included most of the Atlantic and Gulf coastal plains, from the Carolinas to Texas, and much of the Mississippi Valley. Extensive foresting, especially along river bottoms, destroyed the bird's natural habitat and the supply of wood-boring beetles that were its chief food. The last undisputed sighting of an ivory-billed woodpecker in the United States was in 1941. Since that time, there have been occasional rumors of sightings in Louisiana swamps or Texas thickets, but even the rumors have become increasingly rare. The ornithologist Frank Chapman described the call of the ivory-billed woodpecker as "the distant note of a penny trumpet." There are no more penny trumpets in the shops of America, but apparently there are still a few ivory-billed woodpeckers tooting in Cuba. Now that they have been found out the Cuban government has promised to restrict logging in the area where the birds were sighted. It is not known to what extent the government can restrict the influx of ardent birdwatchers intent upon adding this red-crested monarch to their lists.

Supporters of legislation to protect endangered species rightly emphasize the finality of extinction. When something is "dead as a dodo" it is dead indeed. No one expects ever again to see a live dodo or a live passenger pigeon, and a lot of people never expected to see another ivory-billed woodpecker. The ivory-bill is not the only creature that returned from "extinction." The most celebrated case is the coelacanth. For years this remarkable fish was known only from fossils tens or hundreds of millions of years old. The coelacanth is one of the primitive lobe-finned fish from which all vertebrates are descended. It was thought to have become extinct at the same time as the dinosaurs. Then, in 1938, to everyone's surprise, a live coelacanth was caught by fishermen off the coast of Africa. The catch threw the scientific world into a tizzy. It was as if a museum fossil had suddenly come to life. A search of the same waters failed to produce another specimen. It wasn't until 1953 that a second coelacanth was pulled from the sea near the Comoro Islands in the Indian Ocean. Since that time, dozens of the great ichthyoid atavists have been found and studied.

The strange story of the coelacanth supports the always fascinating possibility of creatures returning from the dead. An astonishing number of people in Britain—and elsewhere—are convinced that a sizable monster lives in the cold depths of Scotland's Loch Ness. The most widely held theory is that "Nessie" (as the monster is affectionately called) is a plesiosaur, a seagoing dinosaur that scientists say has been extinct for sixty-three million years. According to Nessie enthusiasts, a plesiosaur has somehow managed to survive in the cold dark waters of Loch Ness, and to support their claim, they point to the coelacanth. Of course, no serious scientist believes a plesiosaur, or anything like it, lives in Loch Ness; for one thing, only ten thousand years ago Loch Ness was filled with ice, not water. Nevertheless, tens of thousands of dollars have been spent trying to acquire definitive evidence for the existence of a plesiosaur in Loch Ness, so far without success, but with a lot of fun.

Everyone loves the thought of a survivor, even if the survivor survives only as science fiction. In a modest sort of way, the ivory-billed woodpeckers observed by ornithologists in the forests of darkest Cuba turned science fiction into science fact. Perhaps the rediscovery of the elusive bird is not so grand a story as the discovery of a plesiosaur in a Scottish lake—but for the ivory-billed woodpecker, it is the best story of all.

If you are looking for the bluebird of happiness it helps to have neighbors with meadows, hedgerows, fruit trees, organic gardens, and nesting boxes designed especially for bluebirds. I have such neighbors. My walk to work each day takes me through conservation land administered by my town's Natural Resources Trust. It would be hard to imagine a habitat more perfectly suited to a bluebird's needs. And in the spring of 1989, for the first time in twenty-four years, I watched bluebirds.

There was a time when bluebirds were as common as robins in my part of New England. Then, especially in the 1950s and '60s, their numbers plummeted. A generation of New Englanders came to maturity without ever seeing one of these gentle, beautiful birds. For some, the "bluebird of happiness" is a mythical creature, invented by Judy Garland, living somewhere over the rainbow. But bluebirds are real enough: electric blue, robin-red-breasted, plump, round-shouldered insect snatchers, flitting through hedgerows or the branches of a pear tree. Who can see a bluebird and not be happy? The naturalist John Burroughs heard its song as "pur-i-ty, pur-i-ty." Others hear "tru-al-ly, tru-al-ly." No one with an ounce of sentimentality in their soul doubts that bluebirds are both pure and true.

What caused their decline? Bluebirds are hole-nesting birds; the replacement of wooden fence posts with metal posts and the pruning of dead wood from orchards eliminated many natural cavities, and starlings and house sparrows displaced bluebirds from the few remaining nests. But then, starting in the mid-1980s, friends of the bluebird set up thousands of nesting boxes all across New England, with holes sized to exclude starlings, and these undoubtedly assisted the bluebird's return. But nesting holes aren't the whole story. Severe winters took their toll on bluebird populations. Pesticides killed insects on which the birds feed. And, of course, at the time of the bluebird's most precipitous decline, DDT was a likely culprit.

I saw a bluebird in my town in 1965, the last for more than two decades. It was the same year I read Rachel Carson's *Silent Spring*, published a few years earlier. Carson's book exposed massive, indiscriminate use of pesticides, especially DDT, and gloomily assessed the consequences for the environment. *Silent Spring* was a trumpet of doom and a clarion call to action. It may be the most influential book ever addressed to a popular audience by a scientist. Rachel Carson was a marine biologist with a bluebird's disposition. She was not by nature a rabble rouser or a public scold. She would rather have been remembered for her sea books, *Under the Sea-Wind, The Sea Around Us*, and *The Edge of the Sea*. She did not choose to become a founder of the environmental movement in this country; the issue of chemical pollution chose her. According to Frank Graham, author of *Since Silent Spring*, the stimulus for action was a letter to Carson from her friend Olga Huckins, who with her husband maintained a private bird sanctuary behind their home in Duxbury, Massachusetts. In 1957 the state of Massachusetts began aerial spraying of deadly pesticides as part of a mosquito control project. The Huckins land was repeatedly sprayed, and many birds were victims. No one asked the Huckinses if they wanted their land bombarded from the air. Those were the days when DDT was sprayed or dusted over half the landscape as a weapon against Dutch elm disease, spruce bud worm, gypsy moths, mosquitoes, and agricultural pests. The more Carson looked into the use of chemical poisons, the more she became alarmed by the potential for ecological disaster. Her book was a brilliant, impassioned call to arms against entrenched interests in government, agriculture, and the chemical industry.

So successful was Carson's call that within months of the book's publication many states and foreign countries issued bans on DDT. In 1957 the U.S. Department of Agriculture sprayed

4.9 million acres with DDT; in 1968 the figure had dropped to zero. Today, DDT is no longer an issue, but huge amounts of other poisons are still dumped into the environment. Like many of the gifts of Janus-faced science, insecticides and herbicides have the potential for good and harm. Every person will have a different assessment of the benefits vs. dangers of pesticides and of the proper level of use for maximum benefit to society. Carson herself was not opposed to all pesticides. She vigorously opposed long-lasting chemicals like DDT. She suggested that society had embarked upon an unthinking chemical binge, and that silent springs were a high price to pay for the unstoppered cornucopia of plenty. One of spring's voices now seldom heard is the bluebird's. The role of chemicals in driving bluebirds toward extinction is debatable, but the coincidence of the bird's rapid decline with the heyday of DDT use is indisputable. Insect-eating birds are easy victims of insecticides. Bluebirds, like loons and ivory-billed woodpeckers, are sacrificed to the economic interests of the planet's dominant species.

Agricultural interests with their eye on the bottom line, and even backyard gardeners, are powerfully inclined to use the full arsenal of chemical weapons to increase yields. Everybody wants picture-perfect lawns, healthy trees, and supermarkets stuffed with abundant produce. Science, and its handmaiden technology, stand ready to provide the apparatus of abundance, and the chemical industry urges us on with "yes, yes, yes." "Pur-i-ty, pur-i-ty," the bluebirds call, a tonic of wildness. And those who are pleased to see these delightful birds return to our neighborhoods can only answer, "Tru-al-ly, tru-al-ly."

CHAPTER 14

Revenge of the Yokyoks

Thoreau, Emerson, and Hawthorne all record in their journals a moment when the shrill whistle of the Fitchburg Railroad intruded upon the tranquillity of the Concord woods. The track of that railroad passed very close to Walden Pond, and Thoreau especially took note of the way the smoke-belching locomotives disrupted his country reveries, drowning out with their iron-wheeled rumblings the cry of the loon and the song of the bluebird. Meanwhile, not far away, other enterprising Americans were building yet more railroads, and canals, water mills, factories, and ingenious machines for weaving cloth and forging iron. Even as Thoreau leaned upon his hoe, listening for the bluebird's wild song ("Pur-i-ty, pur-i-ty"), the Industrial Revolution gathered irresistible momentum, like a great locomotive lurching into speed. By the end of the nineteenth century, Americans had made themselves internationally acknowledged masters of technology. The nature poets of Concord and the mill-masters of Pawtucket and Lowell are two sides of the American character. Since our beginning as a nation, we have had a love-hate relationship with machines. We have unabashedly flung a web of machinery across the land (and even into space), and at the same time we have longed nostalgically for the simple life of unspoiled wilderness.

It is our ambivalence toward machines that causes us to find so much to admire in the Renaissance genius Leonardo da Vinci. Of the artists and scientists of the past, he is the best known and most revered by Americans. His ingenious mechanical inventions were centuries ahead of his time, anticipating the Yankee ingenuity of the Wright brothers, Ford, Edison, and Bell. And his paintings—Mona Lisa, particularly—evoke a tranquil natural beauty. We imagine that in Leonardo's work the lion of technology lays down peaceably with the lamb of nature.

In the spring of 1987 I observed an exhibit of Leonardo's mechanical inventions at the Boston Museum of Science. Twenty-four large working models were on display, based on drawings in Leonardo's notebooks and fashioned of polished wood, gleaming brass, and fabric. They included a clock, a flying machine, a helicopter, a parachute, a paddle-wheel ship, instruments of war, power tools, and scientific instruments. Leonardo saw in his mind's eye and worked out on paper many mechanical ideas that did not come to fruition until our own time. He anticipated manned flight, machine tools, mass production, and many other aspects of contemporary technological civilization. The models in the museum show were the complement of the Mona Lisa—and of the delicate sketches of children's faces and wildflowers that we find in Leonardo's notebooks. In the models we discover a man rapturously in love with machines. Even the machines of war—represented in the museum exhibit by a scaling ladder, gun carriage, and tank—seem more like toys than weapons of terror. These playful models and the Mona Lisa are the two sides of Leonardo as mythic hero: a Renaissance man who lives in harmony with technology and nature, reconciling the discordant poles of the American character.

But wait! There is another Leonardo, a hidden Leonardo who is seldom put on public display. For every delicate wildflower among Leonardo's drawings there are sketches of violent storms, explosions, and turbulence. For every sweet-faced cherub there is an old man's face distorted by anger or fear. For every tranquil Madonna there are images of men and animals locked in mortal combat. And the weapons of war, as we see them in the notebooks, are not toys: We see drawings of spinning scythes dismembering bodies, bombards raining death-dealing fire, and shells exploding in starbursts of decapitating shrapnel. Leonardo's apparently beatific vision of nature and machines was not as harmonious as it sometimes seems. It is true that Leonardo bought caged birds in the shops of Milan that he might set them free. And, yes, among his technical sketches (and the museum models) are many machines designed to increase human well-being and alleviate drudgery. But Leonardo also perceived a dark conflict between nature and technology that resisted resolution. He wanted to learn from nature a more humane way of living, with machines as willing servants, but what he discovered in nature was not always pretty, and his studies did not lead him to a technological utopia. The evidence of the notebooks is conclusive: There is a grim and terrible underside to the genius of Leonardo. His experience offers little hope that we will resolve our own love-hate affair with machines.

Nothing more perfectly exemplifies our ambivalent relationship with technology than atomic energy, and no other technology so starkly illustrates the two-edged blade that reaps and slays, the scythe/sword of Leonardo. Not even the Strategic Defense Initiative (Star Wars) or other mega-technologies of destruction evoke more passionate debate or touch deeper human emotions. Atomic energy is the life force of the stars subdued and channeled by human ingenuity, and plunked down—God forbid!—in our own backyards.

The world's first nuclear power station, at Shippingport, Pennsylvania, came on line in the late 1950s. After the horrors of

Hiroshima and Nagasaki, the Shippingport plant seemed to vindicate our hard-won knowledge of the atom's secrets. Here at last was a benevolent, peacetime use for atomic energy. Magazines of the time were full of articles with titles like "The Atom: Our Obedient Servant." For many of us, it was the dawning of an age bright with promise and plenty. A few critics warned of dangers. There was vague talk of "meltdowns," and explosions, and nuclear waste that would be active for ten thousand years. But all of that seemed a trifle hysterical when weighed against the obvious advantages of nuclear energy over conventional sources of power. Coal- and oil-fired power stations pollute the atmosphere, day by day, and not just at the moment of a hypothetical disaster. Strip-mining of coal and the exploitation of oil reserves devastate huge areas of wilderness. Carbon dioxide emissions from the burning of fossil fuels hold the ominous potential for changing the earth's climate. Our cities and our lungs are dirty with soot. Acid rain (caused by fossil fuel emissions) decimates our forests and the wildlife in our lakes and streams. By contrast, nuclear energy seemed kind to the environment. In the mid-sixties I visited the snow-white Yankee Power Station, the first nuclear generating station in New England. It sat quietly in a green valley in western Massachusetts, spinning out nonpolluting kilowatts and being a good neighbor, unobtrusive and strangely beautiful, a vision of the City of Oz in a Berkshire dale. This, I thought, is the future.

During the next two decades, doubts began to grow. As nuclear reactors proliferated, the problem of radioactive wastes became especially intractable. Accidents at a number of plants instilled doubts about safety. It became increasingly clear that nuclear plants leaked radiation into our water and atmosphere. The partial meltdown at Three Mile Island, in 1979, sounded a death-knell for nuclear energy in America. The catastrophic explosion and fire at Chernobyl in the Soviet Union, in 1987, may have been the funeral. The Department of Energy has predicted tens of thousands of extra cancer deaths over the next fifty years as a result of radioactive fallout from Chernobyl. For many people, this is an intolerable toll of human life, all the worse because the agent of death came invisibly on the wind.

In fact, Chernobyl-related deaths will be a tiny fraction of the 630 million cancer deaths that will occur worldwide during that same fifty-year interval. They may also be a small fraction of the deaths that will occur because of other toxic side-effects of technology, including pollution from conventional power stations. Nuclear power advocates say the Chernobyl statistics should be kept in perspective: that nuclear technology should be made safer, not abandoned. They point to the example of France, where a well-managed nuclear power system generates an abundance of clean, cheap energy. It may be true that nuclear power is indeed a "low-risk, high-dread" technology, but that is little comfort to the people who live near Chernobyl, or in the neighborhoods of existing nuclear stations. The level of risk evidenced by the accidents at Three Mile Island and Chernobyl has become politically unacceptable in this country. No new nuclear plants are likely to be built in the United States for the next decade. And maybe never. Although many nations remain committed to the nuclear option, in America the golden age of nuclear power is over.

Or is it? A new kind of nuclear technology looms on the horizon: power from fusion. Present-day reactors generate energy by splitting apart the nucleus of uranium or plutonium atoms, a process called fission. In fusion technology, energy is produced by fusing together nuclei of deuterium and tritium, both forms of hydrogen. The fuel for fusion—hydrogen—is a component of water. It is cheap, inexhaustible, and available to all. A teaspoon of deuterium has the energy equivalent of three hundred gallons of gasoline. The amount of deuterium in a large swimming pool could supply a major city's electrical needs for a year. Unlike uranium and plutonium, deuterium is not radioactive, and tritium only mildly so (it will eventually be possible to do without tritium). The "ash" of the fusion reaction is the harmless gas helium.

Still, formidable problems remain to be solved before fusion energy becomes a practical reality. For the fusion reaction to occur, the fuel must be raised to a temperature of more than fifty million degrees Celsius. The trick is to contain the fuel while simultaneously heating it to these extreme temperatures. Two approaches show promise: containing the fuel with magnetic fields while heating it electrically; and heating tiny fuel pellets with powerful laser beams. (At the time of this writing, the so-called cold fusion process, which uses metallic crystals to catalyze fusion at room temperature, appears unfeasible.) Success may be decades away, but optimism is growing that commercial fusion power lies within our grasp.

Fusion: An inexhaustible energy resource for the twenty-first century. Clean, safe, cheap. Our Obedient Servant. The language used by the advocates of fusion is disturbingly familiar. The lion of technology and the lamb of nature lie down peaceably together. Swords beaten into plowshares. Will fusion solve our energy problems once and for all? Or will the Janus-faced cycle of promise and disillusionment begin again?

Monstrously expensive nuclear reactors sit idle at Seabrook, New Hampshire, and Shoreham, Long Island. The cost of the Shoreham plant has soared dozens and dozens of times over the original estimate, and now, after decades of planning and construction, some experts believe the best plan is simply to abandon the plant. Citizens of New Hampshire wonder how they will ever pay for the Brobdingnagian monster at Seabrook. The two plants are colossal concrete symbols of the intractable perils of megatechnology. One might also cite the B-1 bomber, a $280-million airplane that is so technologically sophisticated it seems to have a hard time staying in the air (three planes crashed in the testing stage). Or the astronomically expensive Strategic Defense Initiative, a national defense scheme of such overwhelming complexity that many critics believe it can never work.

Rube Goldberg, where are you now that we need you?

Rube Goldberg was our philosopher of technical excess. His loony inventions, widely published in American newspapers between 1914 and 1964, helped keep our technological exaggerations in perspective. He is the only American to have his name become a dictionary word while still alive. A "Rube Goldberg" invention uses a ridiculously complicated mechanism to achieve a simple result. Every engineer or technical planner with responsibility for more than a million dollars of public money should be required to read the complete collection of Goldberg's cartoons.

Consider Goldberg's "Simple Way to Open an Egg Without Dropping It in Your Lap." When you pick up your morning paper an attached string opens the door of a bird cage. The bird exits, follows a row of seeds up a platform, and falls into a pitcher of water. The water splashes onto a flower and makes it grow, pushing up a rod attached by string to the trigger of a pistol. The report of the pistol scares a monkey, who jumps up, hitting his head against a bumper, forcing a razor into the egg. The egg drops into an egg cup, and the loosened shell falls into a saucer.

Or how about "The New Household Collar-Button Finder." A man seeking his lost collar button plays "Home Sweet Home" on the oboe. A homesick goldfish is overcome by sadness and sheds copious tears, which fill the goldfish bowl and cause it to overflow onto a flannel doll. The doll shrinks, pulling a string attached to the power switch of an electromagnet. The magnet attracts an iron dollar attached to a thread, lifting a cloth that covers a still-wet painting of a dog bone. The dog licks the bone, gets sick from the paint, and goes looking for relief. He mistakes the collar button for a pill. When his teeth strike the hard brass button he gives a howl, alerting the man to the location of the missing item.

Goofy inventions like these entertained two generations of Americans, and kept us aware of our propensity for technological overkill. With zany good humor, Goldberg showed us how to laugh at our foolishness. He took a college degree in engineering, but cartooning was his life. His biographer, Peter Marzio, tells us that "despite his constant grumblings against the automatic life, Rube loved complex machinery. He marveled at its labor-saving potential, its rhythm, and even its beauty. But he also treasured simple human values and inefficient human pastimes such as daydreaming and laughing." Like most Americans, Goldberg loved and hated machines. According to Marzio, he was keenly aware of the modern dilemma: How can mechanization be introduced and used without demolishing the life-giving harmony between man and his environment. Goldberg knew that technology grows unwieldy because of our insatiable desire for the very latest inventions, at whatever the cost in money or frustration. He imagined an army of "Yokyoks," tiny green men with long, straight noses and red-and-yellow gloves, who carry an assortment of tools and go about fouling the works—clogging holes in saltshakers, causing pens and faucets to leak, blowing fuses, letting the air out of tires. Goldberg warned us against the "gadget-strewn path of civilization." The more complicated our machines become the more opportunities the Yokyoks have to drive us crazy. Yokyoks presumably love nuclear power stations, B-1 bombers, and Star Wars computers, love the superabundance of opportunity for jamming valves, shorting circuits, causing leaks, and transposing bits and bytes.

If Rube Goldberg hadn't existed, we would have needed a Rube Goldberg to invent him. Certainly, we have no shortage of doomsayers who rail against the invidious influence of technology in our lives. But what made Rube Goldberg's critique of technology effective was his unabashed affection for machines. He did not wish to see the technical apparatus of modern civilization dismantled; he merely wanted to make room between the cogs for a little human fun. He resolves the American technological dilemma with a laugh. His inventions, for all their bizarre exaggeration, served simple human needs, things ignored by the overblown schemes of government and industry. Finding a lost collar button. Removing lint from wool. Getting gravy spots off a vest. Preventing cigar burns on carpets. And Goldberg didn't require megawatts and megabucks to accomplish these tasks. His machines employed rabbits, spaniels, old shoes, firecrackers, banana skins, frogs, string, umbrellas, leaky fountain pens, cheese, and balloons. His technological arsenal was the stuff of basement clutter and yard sales.

Rube Goldberg died in 1970 at the age of eighty-seven. If he were still alive, we could ask him to design a cheaper and more humane replacement for idle nuclear power stations, the B-1 bomber, and the Star Wars defense system. A good laugh at the result might help us find our way out of the quagmire of megatechnology.

CHAPTER 15

The Anticipation of Nature

The C-word is back. "Creation," long taboo, tainted with the odor of mysticism, a pariah concept that dared not speak its name, is back in the vocabulary of science. I am not talking about so-called "Creation science," the activity on the part of Christian fundamentalists to prove the literal truth of Biblical creation, which is not science at all. Rather, I am talking about a new willingness on the part of physicists, astronomers, biologists, and neurologists to address the issue of how the world came to be. They are wresting from philosophers and theologians the biggest questions of all: Why is there something rather than nothing? Are the laws of nature unique or necessary? Is life and mind inevitable? What will be the universe's ultimate fate?

Look at a few titles of recently popular books by scientists: *The Creation*, by physical chemist P. W. Atkins; *The Great Design: Particles, Fields, and Creation*, by physicist Robert K. Adair; *The Cosmic Blueprint: New Discoveries in Nature's Creative Ability to Order the Universe*, by physicist Paul Davies; *The Creation of Matter: The Universe from Beginning to End*, by physicist Harald Fritzsch. The cosmic blueprint! The universe from beginning to end! What kind of chutzpah is this? For centuries scientists prided themselves on the restraint of their questions, and on their distaste for speculation that was not firmly grounded in observation. They scorned "essences," "first causes," and, most disdainfully, any hint of "design." Now it seems no question is too big or too speculative for their vaulting confidence.

No, it's not as bad as all that. The humility of scientists has not become entirely enfeebled. And the new creationists have justifiable grounds for a certain degree of chutzpah.

Physicists and astronomers have pieced together a remarkably consistent picture of how the world began—a "big bang" creation, fifteen or twenty billion years ago, that brought matter, space, and time into existence. The author of *The Great Design*, Robert Adair, says this: "The history of the universe from the first one-thousandth of a second after creation until the present is now a well-documented part of knowledge." It's an audacious claim, but true, or at least as true as any scientific knowledge can be said to be true. Adair is an Associate Director of the Brookhaven National Laboratory on Long Island. At Brookhaven, and at other high-energy accelerator labs, physicists speed up subatomic particles to velocities close to the speed of light, then smash them into each other to see how they behave. Theoretical cosmologists use this knowledge to reconstruct the first moments of creation—the big bang—when the temperature and energy of particles were extremely high. Calculations based on the observed properties of matter at high energy predict the kind of universe that should have emerged from the big bang, and the predictions can be compared with the universe we observe today. How are the galaxies distributed in space? What is the relative abundance of elements in the universe; how much hydrogen, how much helium? What is the mass density of the universe? What is the universe's rate of expansion? Is space filled with neutrinos? Can we observe the "flash" of the big bang? So far, the observations of astronomers are consistent with discoveries of the high-energy physicists, so much so that most scientists now believe we understand in broad outline how the universe came to be—at least since the first "one-thousandth of a second."

Emboldened by success, the high-energy physicists now want to build an even more powerful accelerator—the multibillion-dollar Superconducting Supercollider—that will let them explore the behavior of matter at even higher energies, and therefore at an earlier epoch in the universe's history, closer to time zero. Theoretical physicists are looking for a Grand Unification Theory (GUT), a "theory of everything" that will unite the known forces of nature in one elegant mathematical formulation, as they were united in the first instant of creation. And a new generation of giant telescopes, under construction or currently being deployed, including the trouble-plagued Hubble Space Telescope, will let astronomers peer deeper into space and further back in time to test the predictions of the physicists.

Have the experimental and theoretical successes of the scientists been a prelude to hubris? Have scientists fallen prey to the same cocky, "too-big-for-their-britches" speculations of which they formerly accused philosophers? Quite the contrary, as the reader of the above books will quickly discern. The books are marked by considerable modesty. They reflect a growing respect on the part of physicists for nature's unity, and an almost mystical reverence for the glory of creation, while also stressing the careful interplay of speculation and empirical observation that has always been the basis of good science. If the speculations are grander and more ambitious than they have been in the past, it is because new technologies have made possible wholly new ways of putting nature to the test. Some of the new creationists emphasize the role of chance in the universe; others see the workings of a grand design. Some dismiss traditional religion as irrelevant; others stress the complementarity of science and religion. Some are atheists or agnostics; others are theists. But all of them insist that only observation of nature can teach us how the universe began. All of them admit that our present knowledge of creation is partial and tentative. And where there are still gaps in our knowledge, all of them are willing to say humbly, "We don't yet know."

It has been less than fifty years since Ernest Lawrence was awarded the Nobel Prize in physics for his invention of the cyclotron. Lawrence's first particle-accelerating machine was four inches in diameter and constructed from window glass, brass plate, and sealing wax. It was the sort of thing any clever fellow with a few bucks could build in his basement. But if you want to test current theories for the origin of the universe, you will have to come up with five billion dollars for an accelerator with energies ten million times higher than those achieved by Lawrence with his first machines. The dream machine of today's high-energy particle physicists is the Superconducting Supercollider, now under construction near Waxahachie, Texas. It will be contained in a tunnel fifty miles in diameter and will accelerate particles to energies of twenty trillion electron volts (an electron volt is the amount of energy that could be imparted to a proton by a single flashlight battery). Obviously, this is not the sort of gizmo you build in your basement. Asking nature ultimate questions does not come cheap. If you want to work with the Superconducting Supercollider, you will have to make yourself part of that elite group of high-energy particle physicists who will have exclusive access to the machine.

The Superconducting Supercollider is only the most dramatic example of the tendency in science toward bigger and more expensive instrumentation. The day when a Michael Faraday could discover fundamental laws of electromagnetism with only a few coils of wire and a magnet is past. The day when a rich amateur like James Joule could put together a few paddlewheels and pulleys and verify conservation of energy is gone forever. The frontiers of science today are in the realm of the very big, the very distant, the very small, and the very early: To explore those frontiers requires machines of great sophistication and staggering cost. A glance at the instrument ads in *Science,* the weekly journal of the American Association for the Advancement of Science, tells the story:

- "Discover the Power of the ACAS 470."
- "Think What You Could Do Now, If You Use FPLC."
- "The Sorvall RC-Ultras—A Whole New Way to Save Time."
- "Finally, a Thermospray LC/MS for Less Than $150,000."

If you are an academic researcher and you want an ACAS 470, or FPLC, or an RC-Ultra, or an LC/MS, you will need more money than you are likely to get from the college bursar. And if you don't have one of those devices, the researcher next door is going to get the job done first. In science, priority is everything, and priority means having access to the machine that is fastest, sees farthest, or generates greater energies. What all of this means is that there is real change going on in the way science is done. Research is increasingly dependent upon government funding or the promise of commercial payoffs. The independent genius, like Faraday, Joule, or Lawrence, doesn't have a chance. Today, a paper in the field of high-energy particle physics may have as many as a hundred co-authors, all of whom collaborated in the use of a colossal machine. What any one author contributed is anybody's guess.

Writing in *American Scientist,* Philip Abelson, co-discoverer of the element neptunian, suggests that the cost of research is driving academic researchers into three camps. One group forms the teams that use the mega-facilities of big science—the gigavolt accelerating machines, the multi-meter telescopes, and the supercomputers. They spend considerable time away from the campus and have little contact with students. The second group includes those who have received grants or contracts to do work on campus. They usually supervise a group of graduate students or postdoctoral fellows, but they spend a substantial part of their time managing projects, writing reports, and seeking grants or grant renewals. The third group consists of those whose grant applications have gone unfunded. Often, they are unable to get support from their institutions for even modest research programs.

Some critics believe that the tendency toward big science is not necessary. They concede that a Superconducting Supercollider or a Hubble Space Telescope is glamorous, but they maintain that the money could be more usefully spent by supporting a broader range of research on a more modest scale. Other scientists insist that the highest priority is to explore the frontiers of knowledge, regardless of the cost. If we want answers to such questions as "How did the universe begin?," pure speculation is futile. Observation—of the very small, the highly energetic, and the very far away—is the only way to truth.

As I write, the Hubble Space Telescope has not yet lived up to its promise. Problems with the mirrors degrade the images astronomers had hoped to see. Not the least of the things astronomers had hoped to see with the new instrument are things that cannot be predicted. The same could be said for the Superconducting Supercollider. If the space telescope and the supercollider confirm present theories for the origin and evolution of the universe, that will be satisfying; if they reveal things unanticipated, that's even better. Delight in the unexpected is part of the lifeblood of science. Almost alone among belief systems, science welcomes the disturbingly new.

Putting the space telescope into orbit and building the supercollider will be neither easy nor cheap. It would be nice if we could probe nature's secrets without the risks and expenses associated with such complex instruments. In a little book published some years ago, called *The Anticipation of Nature,* Rom Harre reminds us that "easy ways of doing hard things" have tempted scientists as they have tempted all men from time to time. There is always a recurring hope that we can discover truth about the world without (as Harre calls it) "the tedium of inspecting Nature." For example, Galileo's contemporaries had several "easy ways" of making statements about physical reality. The teachings of Aristotle were considered by many to be a reliable guide to truth about the world. The Scriptures were another trustworthy source of physical knowledge. If Joshua commanded the sun to stand still at the Battle of Jericho, then it could not be doubted that it was the sun that moved, rather than the earth, the theory of Copernicus notwithstanding.

When Galileo turned his new telescope on the heavens he saw things that rocked natural philosophy to its foundations. Not even Galileo could have anticipated the myriad stars of the Milky Way revealed by his instrument, or the mountains and valleys of the moon, or the satellites of Jupiter. Everything he saw reinforced his conviction that Copernicus was right about the motion of the earth and the central stability of the sun. Galileo never tired of pointing out that the evidence of his telescope carried more weight than a literal interpretation of Scriptures or the nonempirical speculations of philosophers. For his efforts on behalf of the senses, the great Florentine physicist is sometimes called the father of empirical science.

It is less widely known that Galileo himself took shortcuts to the truth. Some of his statements about the world were based as much on his gut feeling about the way the world *should be* as on experiment. Many of Galileo's shortcuts were inspired, and met the test of time. Others led him down blind alleys. For example, Galileo's belief that he had observed seas on the moon was the product of wishful thinking, and was later disproved by closer inspection. Even the greatest scientists sometimes fall victim to falsely anticipating nature. When the equations of Einstein's theory of gravity (general relativity) predicted that the universe should expand or contract, he gratuitously added an extra mathematical factor to suppress the instability. It was inconceivable to Einstein that the universe should be other than steady and eternal. The conviction of stability came crashing down in 1929, when Edwin Hubble announced startling observations made with the new hundred-inch telescope on Mount Wilson in California: The galaxies were in fact rushing outward. Einstein immediately went to California to confer with Hubble. With his wife, Elsa, he was given a tour of the observatory, and it was explained to them how the huge instrument was used for determining the structure of the universe. "Well, well," said Elsa, "my husband does that on the back of an old envelope," and indeed it is part of the mystery of mind and nature that the backs of envelopes can be vehicles of discovery. But Einstein knew that the evidence of observation is the final arbiter of truth; the Mount Wilson telescope, the largest in the world at that time, took precedence over his instinctive feeling for how the world should be. He quickly admitted his error and deleted the offending factor from his theory.

Like Galileo's instrument and the Mount Wilson telescope, a properly functioning Hubble Space Telescope or Superconducting Supercollider will almost certainly reveal things about the universe that run counter to some of our most cherished beliefs. Scientists will be pleased if the new instruments confirm their speculative theories, but they will be disappointed if there are not a few unsettling surprises. It is the glory of science that part of its value system is a willingness to accept the unacceptable when nature instructs us that it is time to do so. Scientific theories are human constructions. Some theories have a generality or a beauty that makes them seem irresistibly true: Einstein's conviction of static galaxies is an example; another is Galileo's anticipation of a watery, earthlike moon. But all past attempts to make generality or beauty the sole warranty of truth have failed.

The "tedium of inspection" is essential; thus, the requirement of always new and more sophisticated instrumentation. As Rom Harre says, "the anticipation of Nature is a fraud." The complexity and cost of instruments must sometimes be in proportion to the boldness of our questions.

CHAPTER 16

Down on the Paluxy River

Down on the Paluxy River in Texas there are human footprints in a stratum of sedimentary rock that bears the tracks of dinosaurs. Or so claim adherents of some fundamentalist religious groups. The Paluxy tracks, and similar markings at other sites in the American west, have become one of the centerpieces of the fundamentalist anti-evolution crusade. For a decade, the purported "man tracks" in Texas have been touted by creationists as proof, once and for all, of the falsity of evolution.

The Paluxy River sedimentary strata date from the Cretaceous Period of geologic history, 120 million years before the present. Creationists claim that the rocks are only thousands of years old, and, they say, the fossil footprints show that humans and dinosaurs coexisted, most likely in the time preceding the flood of Noah. But, of course, there are no human footprints in the Cretaceous rocks of Texas, or any other rocks that date from the time of the dinosaurs. The purported man tracks have been carefully examined by geologists. Some are not tracks at all, but only erosion features typical of river beds. Other poorly defined "human" footprints are similar to dinosaur tracks in size, pace, and step angles, and are very likely poorly preserved dinosaur prints. So devastating has been the scientific critique that some creationists have begun to hedge their bets regarding a human origin for the controversial tracks.

None of this would be worth talking about if it were not for the fact that science is under a growing attack by people who take things like the Paluxy River "man tracks" seriously. Fundamentalists have urged state legislatures to mandate the teaching of "creation science" in public schools along with evolution, with some success. Such laws have been judged unconstitutional by the courts; but, in spite of setbacks, religious groups maintain pressure on school boards, state legislatures, and textbook publishers to include "creation science" in school curricula, or, failing that, to outlaw the teaching of evolution. Freedom of religion is not at issue. People have the right to believe what they want about how and when the world was made. At issue is whether there is such a thing as *creation science*. At issue is the definition of science itself.

Science is not just a collection of true statements about the world. If I say "the earth is 4.5 billion years old," or "humans evolved from lower orders of life," I have made a scientific statement, but such a statement is not itself science. Science is not *what* we know; science is a *way of knowing*. Observation of nature is the most important criterion for the validity of scientific truth, but it is not the only criterion. Different people can interpret the same observations differently (the Paluxy River "man tracks" are a case in point). An equally important criterion for truth is consistency. What we hold to be true in one area of science must not contradict what is held to be true in another area. Science is not a smorgasbord of truths from which we can pick and choose. A better image for science is a spider's web. Confidence in any one strand of the web is maintained by the tension and resiliency of the entire web. If one strand of the web is broken, a certain relaxation of tension is felt throughout the web. We believe in the evolution of life through geologic time because of what we have learned in many other areas of science. Our confidence is assured by the success of the entire ensemble of truths. Scientific truths are tentative and partial, and subject to continual revision and refinement, but as we tinker with truth in science—amending here, augmenting there—we always keep our ear attuned to the timbre of the web.

Fundamentalists use film, video, electronics, computers, and communication satellites to spread the anti-evolution message; indeed, they have become masters of the electronic media. Ironically, these technologies are based on the same ensemble of physical principles that lead us to believe that life evolved over eons of geologic time. You can't tear down one part of the web of science unless you are willing and able to rebuild a structure of understanding that works as well or better than the one you have disassembled. This is what the creationists are unable to do. They pick and choose their scientific truths, ignoring the web of reinforcing relationships. And that is why there is no such thing as creation science.

There is more at risk in the anti-evolution crusade than a particular view of the origin of the world. At risk is the ability of the next generation of Americans to distinguish science from nonscience. Science is confidence in the human mind's ability to discover some measure of truth about the world. Science is humility in the face of nature's complexity. And above all, science is a respect for consistency as a hallmark of truth. In at least one thing the creationists are right: If humans walked with dinosaurs, then evolution is false. And by the same test, much of what we know about geology, astronomy, physics, chemistry, biology, and medicine can be thrown out too.

Anti-evolutionists often claim that the theory of evolution is not scientific even by the standards of science. No hypothesis (they say) can qualify as science unless it can be tested by controlled experiment, and evolutionary happenings, which occurred at distant times in the past, are unique, unrepeatable, and irreversible. Darwin and his followers may claim that fishes turned into amphibians and reptiles into birds, but there is no way the theory can be put to the experimental test.

As paleontologist Stephen Jay Gould has pointed out, the fundamentalist claim that evolutionary theory is unscientific arises from a wrong notion of scientific method. There is no such thing as *the* scientific method. There are many methods by which science seeks truth, and one of them is the historical method. Like the theory of continental drift or the theory of galaxy formation, the theory of evolution is a historical science. It describes events that occurred in the past, on a time scale that is long compared to human experience. We are confident that evolution occurred for the same reason we know that Columbus discovered America in 1492. It is a matter of rigorous inference from historical data. In the case of evolution, the data are the fossils in the rocks. Gould believes that Darwin's chief contribution to science was not the theory of evolution itself, but a convincing demonstration of the usefulness of historical methods in science.

Although there is clearly no way we can recapitulate billions of years of evolution in the laboratory, there are lots of ways *within the historical method* for testing hypotheses. One way is to create computer models of historical events. Events that require millions of years in the real world can be simulated with computers in a modest amount of time. Karl J. Niklas and his colleagues at Cornell University have reported in *Scientific American* a fascinating exercise in computer modeling of biological evolution. They used a computer to study the early evolution of land plants. Millions of years of variation and natural selection unfold in minutes on the screen of their machine. Primitive plants stand up, spread branches, scatter spores, and reach for the sun—all electronically.

First, the Cornell researchers formulated hypotheses about what factors had the greatest effect on plant evolution. They selected the ability of the plants to gather sunlight for photosynthesis, to support vertical branching structures, and to disseminate seeds or spores. The efficiency of all of these things were defined in simple geometrical ways. The plants in the computer model included only simple "stick-figure" branching structures. The probability of branching, the branching angles, and the rotation of the branches about the trunks or stems were controlled by letting the computer simulate random "genetic" mutations of individuals within a species. The program maintained a degree of "genetic" continuity between ancestors and descendants.

If nature is red in tooth and claw, so was the computer. Niklas and his colleagues let the computer play war games. On the field of battle at the start were only simple ground-hugging plants, like those that appear earliest in the fossil record. The computer cranked out successive generations of plants, allowing for mutations, and scored each new species on its ability to gather sunlight, disseminate spores, and avoid the shade of its neighbors. The winners continued into the next generation. The losers were eliminated. The plants that won the war games look astonishingly like modern trees. And the forms of the intermediate computer plants are consistent with the fossil record for plant evolution. The Cornell researchers do not claim that the computer games *prove* the hypotheses of evolution by genetic mutation and natural selection, only that they raise our confidence in the usefulness of the evolutionary hypothesis to explain certain historic events that occurred hundreds of millions of years ago. Opponents of evolution will object that the Cornell computer model is an oversimplification of a very complex set of circumstances occurring in nature. Niklas readily admits this. But even a glance at the lovely plants unfolding on the screen of his computer impresses one with the power of the method.

We all use maps for travel or reference that incorporate only a few features of the terrain they describe. Every map is a simplification of a real landscape; nevertheless, maps are enormously helpful, and it is hard to imagine how we could get along without them. Science is a map of reality. Evolution is a map of the historical past. As a map, it is partial and incomplete, but far more useful than any alternative map that has yet been devised for explaining the vast catalog of fossil evidence that is documented in stone.

It is astonishing to learn that, according to pollsters, 86 percent of Americans believe that creationism should be taught in public-school science classes, either alone or along with evolution. Clearly, scientific criteria for truth are not widely accepted, or generally understood. And even if scientific criteria are granted, the religious fundamentalist has an unbeatable ace up his sleeve: miracles. God has the option of suspending the laws of nature at any time; therefore, science, which deals only with natural law, cannot prove or disprove fundamentalist claims for any supernatural or miraculous event. If *creation science* is found lacking by the standards of science, then the fundamentalist can always fall back on *creation miracle*.

The famous Shroud of Turin offers a superb occasion for examining the relationship of science and miracles. The shroud is a linen cloth preserved in the cathedral at Turin, Italy, bearing the likeness of a man and purporting to be the winding sheet of Christ. By now nearly everyone has heard the result of the carbon-14 dating tests on the shroud, which indicate a medieval age for the cloth. The official report of the investigating scientists appeared in the British journal *Nature* early in 1989. The report is in many ways more interesting than the result of the tests. It is a classic illustration of the scientific way of knowing.

Question: How old is a certain piece of cloth? Forget for the moment what makes this particular piece of cloth historically interesting; can we determine its age? The answer: Yes, by a method known as carbon-14 dating. Carbon dating exploits the precisely measurable radioactive decay of carbon 14 atoms. The details of how the method works are not—in this context—important. The point is that there are scientists who claim they can find out how old something is by counting atoms. Can we believe them? Can we trust their results? Let's put them to the test. Take wood from the heart of a one thousand-year-old sequoia tree whose age can be determined by counting rings. Or a splinter from a beam in Paul Revere's house. Or a thread from your great-great-grandmother's wedding dress. Send the sample to a carbon-dating lab and ask them to determine the age. Don't tell them you already know the answer, and see if they get it right. Now, take my word for it, this has been done, over and over again. Carbon dating has been thoroughly tested and calibrated with objects of known age. The method is complicated—it is based on some high-powered chemistry and physics and expensive equipment—but it works, invariably. Any skeptic who can afford the fee can put it to the test.

Three carbon-dating labs, in Zurich, Oxford, and Arizona, participated in the test on the Shroud of Turin. Along with samples from the shroud, each group was given three control samples of cloth—linen from a nine hundred-year-old Nubian tomb, linen from a second-century A.D. mummy of Cleopatra, and threads from an eight hundred-year-old garment of St. Louis d'Anjou. None of the samples were identified. None of the labs communicated with each other until the results were in. The final report of the investigation describes in detail the methods by which samples were taken from the shroud, then cleaned and prepared for analysis by laboratory staffs. Also described is a statistical analysis of the results. All three labs agreed on the ages of all four samples within experimental error, and all three labs correctly dated the control samples. Conclusion: The Shroud of Turin is medieval. It does not date from the time of Christ.

Is this conclusion absolute? Of course not. No scientific test can prove anything with absolute certainty. As one of the Oxford investigators wrote (in a letter to *Nature*), "If we accept a scientific result, we must exercise a critical notion of the probabilities involved. If we demand absolute certainty, we shall have to rely on faith." Science is not faith. Science is common sense. There is always a remote possibility that some unknown agency acted on the shroud in such a way as to cause the carbon dating experiments to give a false result. For example, a Harvard University physicist has proposed (whether seriously or not, I do not know) that neutron emission from the radiant body of the resurrected Christ might have transformed some nonradioactive carbon into carbon 14 by neutron capture, thus invalidating an important assumption upon which the age measurement is based. The three laboratories apparently considered this bizarre possibility, and their reply (articulated by R. E. M. Hedges of Oxford) goes something like this: (1) No known physical process could produce a neutron flux of the required magnitude, and if a supernatural explanation is proposed, then there is hardly any point in doing the experiments. (2) It would be an amazing coincidence if the proposed neutron emission was such as to make the apparent age of the cloth coincide with the very time—mid-fourteenth century—when the shroud is first mentioned in history. (3) Science cannot disprove supposed miracles. What science can do is confirm a perfectly natural explanation for this remarkable object which is consistent with everything else we know about the world. Namely, the shroud is a fourteenth-century icon or forgery.

It is to the credit of Church officials in Italy that they authorized the carbon-14 tests and accept the results. Their actions are consistent with a recent declaration of Pope John Paul II on the proper relationship of science and theology. "Science can purify religion from error and superstition," wrote the Pope, "[and] religion can purify science from idolatry and false absolutes." He continued: "The unprecedented opportunity we have today is for a common interactive relationship in which each discipline retains its integrity and yet is radically open to the discoveries and insights of the other."

Scientists are skeptical of miracles, because most (or all) purported miracles, like the Shroud of Turin, are consistent with natural explanations. Nevertheless, many people will still want to believe that the Shroud of Turin is the winding sheet of Christ, or that the world is only a few thousand years old, and that is their right; no carbon-14 tests, or any other scientific test or computer simulation will dissuade them. For one reason or another, they prefer the apparent certainties of miracles to the considered probabilities of natural science.

CHAPTER 17

An Appetite for Baloney

They are "the cosmically charged cornerstones upon which the great pyramids of Egypt were built." They are "natural superconductors through which a universe of enlightenment passed to the lost continent of Atlantis." They are crystals, and if we know how to use them they can make us healthy, wealthy, and wise.

Or so believe a growing number of "New Age" people, who use crystals in every aspect of their daily lives, often along with horoscopes, herbs, incense, and elixirs. Beryl, for example, relieves stress. Diamond guards against envy. Azurite heightens dreams. Garnet improves self-esteem. And good old white quartz is an energizer, channeling into our environment the very best cosmic vibrations. Bookstores are jam-packed with crystal self-help books, or you can pay seventy-five dollars an hour to visit a crystal consultant. We knew crystals had made it big when a cluster of violet-tinged quartz prisms appeared on the cover of *Time* magazine, in the hands of New Age guru Shirley MacLaine.

Now it so happens that I have on my desk a cluster of quartz crystals almost identical to those on the *Time* cover. And, as a matter of fact, a number of good things began happening in my life at about the time I got the crystals. So have I become a New Age believer? Not likely. Stephen Jay Gould once wrote that "the human mind delights in finding pattern—so much so that we often mistake coincidence . . . for profound meaning." Was the turn in my luck that occurred at the time I obtained the crystals coincidence, or was it cosmic energy channeled through parallelepipeds of silicon dioxide? Science is one highly successful apparatus for winnowing the wheat of profound meaning from the chaff of coincidence. And in the matter of crystals, I'll stick with science.

The coincidence between my crystals and my good fortune was just that—coincidence. Yet I keep the crystals on my desk for a profound reason, and the reason is certainly cosmic. Crystals are windows on the world of atoms. Perhaps better than anything else in our natural environment they exemplify the patterns of order that are built into all of nature on the atomic scale.

The quantitative science of crystals began in the seventeenth century when the Danish natural philosopher Nicolaus Steno noticed that the faces of quartz crystals always meet at the same angle. No matter how big or small the crystal, or its shape, or its place of origin, the angles are the same. As I was writing this essay, I put my cluster of quartz crystals under a microscope and entered a world of spectacular architectural forms—gorgeous, icy-smooth planes colliding this way and that to form what appeared to be a space-age city of towering pillars and plunging chasms. At every edge of every column or canyon, I observed Steno's unvarying angle of one hundred twenty degrees: a hint of constancy in the midst of chaos, a hint of the ordering principles that brings cosmic harmony out of the primeval dishevelment of creation.

In the late eighteenth century, the mineralogist Rene Just Hauy made another remarkable discovery about the nature of crystals. According to the story, Hauy was examining a large calcite crystal in the home of a friend when he dropped it. To his embarrassment, it shattered. When he stooped to pick up the pieces he noticed that every fragment, large and small, had the same basic shape as the original crystal. *"Tout est trouvé!"* shouted Hauy. "All is discovered!" He rushed back to his lab and smashed a calcite crystal into tiny bits, and every piece resembled the parent. Hauy reasoned that the smallest units of the crystal—perhaps even on the atomic scale—would have the same shape. Twentieth-century X-ray technology confirmed Hauy's conclusion. The shape of a crystal is determined by the way atoms of that substance bind together to form "lattices," repeating arrays of space-filling atomic units. In all of nature there are only a few classes of fundamental crystal shapes, and these can be predicted mathematically. Yet no two crystals are exactly alike, and that too is part of their attraction. Tiny traces of impurities add color to crystals: emerald, sapphire, ruby, aquamarine. Pure quartz is clear; the amethyst color of my quartz cluster probably derives from tiny traces of iron impurities, and this delicate departure from perfection enhances the beauty and value of the crystals.

Under the microscope my apparently regular quartz crystals become a showcase of imperfections: color variations, striations, cracks and inclusions of tiny crystals of another order, including some wonderful starburst crystals of goethite (if my guess is right) embedded within the quartz. Patterns within patterns within patterns. All of this beauty hints at things still beyond the crystallographer's ken. Most mysterious of all is the riddle of crystal growth. If crystals grow by adding atoms at the surface, and if atoms fall into place essentially at random, then why aren't those shimmering surfaces that I see in my microscope bumpier? Consider the perfect symmetry of the snowflake: How do atoms attaching themselves at one tip of the growing six-sided crystal know what's happening at another tip? Scientists are not yet sure, but the answer may have something to do with vibrations within the crystal, vibrations of an exquisite sensitivity that maintain a delicate—dare I say cosmic—balance between order and disorder. We are faced here with the deepest and most profound mystery of creation, what the Greeks called the problem of the One and the Many: How does order and beauty emerge from chaos? And this is the lesson of the crystals: *Nature is delicately balanced on a knife edge between the tedium of perfection and the bedlam of randomness*. Beauty, life, and mind emerge on the blade of that fine balance.

Vibration is undoubtedly part of the creative force of the cosmos; pulsars, snowflakes, and hydrogen atoms vibrate, filling the universe with scintillant radiation. And do the crystals on my desk channel those cosmic vibrations into my personal life? I think not. Crystals are delightful icons of design in nature, but the New Age "wheat" of profound meaning is my chaff of coincidence.

The nineteenth-century physicist Michael Faraday said: "Nothing is too wonderful to be true." With that in mind, science must be open to the possibility that an apparently offbeat idea contains the seed of truth. At the same time, it is right to insist that certain evidential criteria be met for an idea to qualify as science. Science is public knowledge. It is based upon observations that can be repeated by any investigator, believer or skeptic, with uniform results. For example, scientists may vigorously debate the agency or mode of evolution, but the fossil record in the rocks is there for anyone to see. The "phenomena" of the parasciences, including crystal therapy, astrology, parapsychology, extra-sensory perception, UFO-ology, and a host of other bogus shortcuts to truth, are generally discernible only to advocates of those disciplines. If science is open to every private vision of reality, then its usefulness as public knowledge is severely impaired. When gurus of magic crystals or astrologers can produce a single convincing experimental test of their claims, one that can be repeated by skeptics as well as believers, only then can they lay claim to legitimate public knowledge.

Astrologically, I'm a Virgo. My reference book on astrology says that Virgos are practical, hardworking, analytical, meticulous, tidy, and modest. That's me, all right, except maybe for the "modest." Coincidence, or profound meaning? What can science tell me about the truth of astrology? There have been many published experimental "tests" of astrology. Those that seem to confirm astrology have been dismissed by scientists on the basis of poor experimental technique. Those that seem to show that astrology is false have been dismissed by astrologers as theoretically biased. Physicist Shawn Carlson of the University of California at Berkeley set out to make a definitive test of the predictive powers of astrology. He tried to devise a test that would meet the most stringent criteria of both the scientific and astrological communities, choosing as advisors for the experiment only scientists and astrologers that were held in high esteem by their respective communities. The astrologers helped formulate what Carlson calls the "fundamental thesis of natal astrology": That the positions of the planets, sun, and moon at the moment of birth can be used to determine a subject's personality traits and tendencies in temperament and behavior.

The experiment consisted of two parts. In the first part, volunteers provided information from which natal charts and interpretations were constructed by professional astrologers. Each subject then attempted to select his own interpretation from a group that included his own and two others chosen at random. In the second part of the experiment, the astrologers were given the natal chart of a random subject and an assessment of that subject's traits based on the California Personality Inventory (CPI). They were also given two other CPIs chosen at random from among all the subjects. The astrologers were asked to pick the personality assessment that best matched the natal chart.

The scientific thesis—that a correct match of subject and natal chart would be purely a matter of chance—predicted a success rate of one-third. The astrologers predicted that their interpretations would be correctly matched with subjects at least half of the time. I will not bore you with an account of the many procedures employed by Carlson to insure that biases on the part of the astrologers, subjects, or researchers did not affect the outcome of the experiment. And can you guess the outcome? The number of correct matches was statistically indistinguishable from a purely random result. You or I could have done as well as the professional astrologers by pulling horoscopes out of a hat. Conclusion: Astrology is baloney.

Scientists, hearing of Carlson's experiment, will say "ho-hum, what else is new?" And believers in astrology will certainly not be dissuaded from their faith by a statistical experiment; their belief in the determining power of the stars was essentially irrational to begin with. As I read the summary of Carlson's research, I had the sense that he would not have been unhappy if the results had come out in favor of astrology. Certainly, he bent over backwards to give the astrologers everything they wanted. It was the stars, presumably, that refused to cooperate. The fact is this: Many people have a greater appetite for superstition than for reality. Every morning, millions of readers of newspapers will turn to the horoscope to see what the stars of the astrologer have to say about their day. Meanwhile, the real stars, those gorgeous nighttime presences ("night's candles," Shakespeare called them), go mostly unnoticed. The stars of the astronomer, brilliant and furious with thermonuclear light, profound beacons of meaning in the universe, play second fiddle to the phony newsprint stars of the astrologers.

I looked up my own horoscope in the newspaper this morning. It suggested that I should be loyal to my mate in any quarrel that might arise with someone else. That's good advice. But, if I am a reasonably mature sort of fellow, I didn't need the stars to tell me that.

A couple of years ago, I used my science column in *The Boston Globe* to debunk astrology. I pointed out the complete absence of any reproducible, empirical evidence linking individual human lives to the stars. I stressed the positive virtues of scientific skepticism, and suggested that astrology is fundamentally anti-rational. For most of my audience, I was preaching to the converted. For the rest—believers in astrology who just happen to read the science pages of the newspaper—no amount of debunking would have made any difference. But one good thing came from the exercise. I heard from a number of professional astrologers and I was impressed by their sincerity. These people were not charlatans. They were convinced of the validity of their craft and motivated by an unselfish desire to help others. They offered gentle responses to my not-so-gentle criticisms. I entered into a brief correspondence with one astrologer, a person who has published several best-selling books on the subject. I even read his books. They were lively, well-written, and fun. As self-help books, this particular author's works contained much good, sensible advice. They evoked a sense of wonder for nature, a positive attitude toward people, and (perhaps paradoxically) a healthy sense of personal responsibility.

But I learned nothing from the books, or from my encounters with astrologers, to persuade me that astrology is anything more than a silly superstition. What I did learn is that there is no way for a scientist to convince an astrologer that his craft is superstitious, and no way an astrologer will convert the scientist. It is not so much a matter of evidence (Carlson's experiments, for example), as an attitude toward evidence. The astrologer and the scientist have different criteria for truth, and, consequently, little hope for resolving their differences. When a scientist sees a person ordering his life by the stars, he sees the surrender of reason. When an astrologer sees a scientist "debunking" his craft, he sees bias, conspiracy, or blinkered fixation on dogma.

A correspondent to the *Globe* took issue with my anti-astrology column this way: "Raymo's argument against astrology is the usual one: astrology can be done away with by simply declaring it irrational. In other words, if we cannot understand why it works, it must not work. The same flawed argument could be used against electromagnetism, particle physics and the force of gravity, with equally senseless results." And it is true. I don't understand in any ultimate sense why electromagnetism, particle physics, or gravity works. Nobody does. But the point is this: Electromagnetism, particle physics, and gravity *do work,* in a way that astrology does not. Experiments of the most exquisite sensitivity can be devised to test the former theories, experiments that can be performed consistently by believers and skeptics alike. Radio communication, nuclear power, and the space program are testaments to the fact that electromagnetism, particle physics, and gravity work. On the other hand, every objective test of astrology that I know of (with the possible exception of one highly controversial study) has resulted in failure. Then why do astrologers continue to insist that astrology works? Here's why. I once had my birth chart done by an astrologer. She labored long over ephemerides and graphs, and then told me I was sensitive, intelligent, basically generous, but sometimes self-indulgent, inclined toward optimism, but subject to occasional bouts of depression. Wow! Who would object to that description of his personality?

In spite of reliance on numbers, graphs, and even computers, astrology has nothing to do with science, and no objective experiments will dissuade an astrologer from his faith. So why raise the subject at all? At the risk of condescending to believers in astrology, let me suggest that there are a few things we can learn from each other. Scientists should ask themselves what is behind the incredible popular appeal of astrology, crystal therapy, and other parasciences. Science is widely perceived as cold and aloof from human emotional needs and aspirations. Until scientists can find ways to more effectively communicate the human face of science, and the ennobling vision of the universe revealed by science, there will continue to be a wide audience for pseudoscience and superstition. And astrologers should ask themselves if they really want to live in a world that is ruled by a slipshod attitude toward evidence. We are frequently reminded by astrologers that such great scientists as Kepler and Newton believed in astrology. True enough. But astrologers should also recall that Kepler's mother was very nearly burned as a witch, and that Newton's university was closed because of the plague. Witchcraft flourished because people imagined causal connections where none existed, and plague vanished from Europe when people took note of causal connections (for example, the connection between rats and disease) that could be verified empirically. Verifiable, repeatable tests of causal connectivity was at the heart of the Scientific Revolution. It is no coincidence that witchcraft and plague disappeared from the western world at the same time that astrology was finally discarded from science.

CHAPTER 18

Of Dragons and Hippogriffs

They just won't go away. They hang around up there, year after year, in their saucer-shaped craft, playing tag with airliners and causing inexplicable blips on radar screens. They love to show off, flying in tight formation, or doing ninety-degree turns at twice the speed of sound. And every now and then, when they get really bored, they land on the surface and treat some lucky (unlucky?) human to a ride. Maybe even whisk him away on a quick trip back to the home planet. These extraterrestrial aliens in UFOs have made themselves a permanent part of our culture.

On August 12, 1986, thousands of people in the eastern United States witnessed a spectacular unidentified flying object. One observer saw "a glowing, spiral pinwheel, standing on end and moving on a line from southeast to northwest." Others reported a luminous disk larger than the moon, with a starlike core. None of the witnesses, including many amateur astronomers, had ever seen anything like it. Police departments and radio stations were swamped with calls. Then, on November 17, the pilot of a Japan Air Lines cargo jet flying over the Arctic Ocean reported being followed for nearly an hour by two strands of lights and a huge "mother ship." The large object was the "size of two battleships," the pilot said, and appeared to be made by a "high technology and intelligence." Blips on a ground-based radar screen seemed to confirm that a craft of unknown origin was in the vicinity of the jet. These two spectacular encounters with UFOs provoked a lively recurrence of interest in spaceships from other worlds. Investigations of the two reports are now complete, but before I reveal how two UFOs became IFOs let me put the sightings into context.

It all started on Tuesday, June 24, 1947. Businessman Kenneth Arnold was flying his private plane above the Cascade Mountains in Washington. Nine circular objects, in tight, diagonal formation, passed within twenty-five miles of his plane. Later, Arnold told a reporter that the objects flew "like a saucer would if you skipped it across the water." The next day all America heard about the flying saucers. Within a month, saucers had been reported from every state in the Union and half the countries in the world. I lived through the excitement. I was twelve years old in January of 1948 when Captain Thomas Mantell, in a P-51 Mustang, chased a saucer up to twenty thousand feet. He lost consciousness and nose-dived into the ground. I remember a headline something like this: AIR FORCE PILOT KILLED CHASING UFO. Big news for a twelve year old. For the next six years I read everything I could find about flying saucers. And there was plenty to read: books and magazine articles by the dozens, newsletters of UFO societies, and the official report of an Air Force UFO investigation called Project Blue Book.

And so began the cult of the UFOs. The cult endures today, as vigorous as ever. I am often asked if I believe in UFOs. The answer is yes. I have seen several UFOs. Anybody who regularly watches the sky is sure to see an occasional unidentified flying object. I remember one night in particular when a bunch of us were standing around in a misty field with a telescope. This thing zipped across the sky from east to west, turned around, and zipped back. Too fast for a plane. Meteorites don't turn around. Whatever it was was unidentified. And flying. An honest-to-goodness UFO. But did it have an extraterrestrial origin? I think not. Most UFOs turn out to have more mundane explanations, and the rest remain simply unexplained.

The spectacular object that appeared over the eastern United States on August 12, 1986, was a cloud of fuel vented from a Japanese satellite launch vehicle, in orbit high above the earth. Similar clouds have been observed in South America from Soviet launchings from Plesetsk, and in Australia from American launchings from Cape Canaveral. They occur at a particular place in the launch trajectory. The Japanese rocket test was the first of its kind, and the cloud of vented fuel the first to appear over the United States. It was the unfamiliarity of the effect that caused all the excitement. What the pilot of the Japan Air Lines jet saw on November 17 was the planet Jupiter, which was very bright at that time and in the same part of the sky as the observed UFO. The Federal Aviation Agency was unable to confirm the sighting of a flying object. A United Airlines pilot in the vicinity of the JAL plane saw nothing. The blips on the radar screen that seemed to confirm the UFO turned out to be "split-radar returns," shadows of the plane's primary echo.

Of the thousands of UFOs that have been reported over the past forty years, not one has passed scientific muster for an object of extraterrestrial origin. But still the cult of the UFO endures, I suspect because visits to earth by extraterrestrials would affirm that the universe is not indifferent to our existence. Cultists will not be dissuaded by talk of vented rocket fuel and "split-radar" echoes. They will say that once again conspiratorial scientists have "explained away" something that doesn't fit accepted theories. It's the same old argument rolled out by the astrologers and crystal gurus: Science is a closed, self-perpetuating establishment that refuses to admit the possibility of unorthodox phenomena. This critique by the UFO-ologists is bankrupt. I won't speak for other scientists, but inside this typical scientific skeptic there is a twelve-year-old boy who wants desperately to believe in the visitors from outer space. He's still waiting for the evidence.

Humans have an appetite for the fabulous. Once that appetite was satisfied by unicorns, hippogriffs, mermaids, or monsters. Today, more often than not, it is satisfied by UFOs, abominable snowmen, Loch Ness monsters, and other pseudoscientific phenomena. I am always astonished that people are interested in such things as astrology, crystal therapy, ESP, and "ancient astronauts" when the real wonders of the world lie all around us, wonders more fabulous than anything pseudoscientists can dream. What Loch Ness monster is more mysterious than the gargantuan black hole that astronomers have detected at the center of the Milky Way Galaxy, a world-devouring singularity with a mass of a million stars? What astrological influence of the stars is more magical than the dazzling dance of the DNA as it goes about its business of making us what we are?

My early, unskeptical enthusiasm for UFOs was probably good training for becoming a scientist. A good scientist must be able to conceive of the fabulous. The historian Lewis Mumford wrote, "If man had not encountered dragons and hippogriffs in dream, he might never have conceived of the atom." According to Mumford, it was from the experience of the fabulous in dreams that humans came to believe that there is more to reality than meets the eye. Dreams gave sleepers access to an unseen world veiled from our senses and from daily experience, but as apparently real as the food we eat. Dreams practiced the imagination.

We no longer believe in the literal reality of dream images, but belief in an unseen world veiled from our senses and from daily experience is an important part of modern science. We believe, for instance, in atoms. We believe in black holes and the big bang. We believe in dinosaurs. These inventions of science are as real to us as were dragons and hippogriffs to our ancestors, and no less fabulous. Consider the straightforward phenomenon of a 747 jumbo jet taking to the air. I understand the physics of flight. I can use Bernoulli's equation to calculate the pressure of moving air on the top and bottom surfaces of the wings, and I know that the pounds-per-square-inch difference is enough to hold up the plane. I am fully acquainted with the natural laws that explain flight. But every time I enter a 747 I'm convinced it will never leave the ground. That it does is a thing as fabulous as any hippogriff.

Or genes. In every cell of my body there is a complete blueprint for making another me, stored on DNA molecules. I've seen electron microphotographs of those cobwebby DNA structures in cell nuclei. The total length of DNA in each cell is about as long as my arm, three feet of genetic "string" stored in each of trillions of individually invisible cells. If you unraveled my body like a ball of twine there would be enough DNA to reach to the moon and back ten thousand times. I can work this out on paper, but it still seems miraculous. To conceive of the DNA, it helps to have practiced on UFOs.

And the most fabulous thing of all is radio. I have a little multiband radio receiver in my office. By poking a few buttons I can listen to the Berlin Philharmonic play Beethoven, or hear a Russian commentator read an English version of the news, or join a BBC guide on a tour of Canterbury Cathedral, or, when conditions are right, tune in to pop music from Australia. And that's in addition to a lively selection of broadcasts from stations in my neighborhood. We live in a sea of radio waves. Right now, Mozart piano concertos and Scott Joplin rags are coursing through my body. I share my office with the voices of Bruce Springsteen, Madonna, Hank Williams, Frank Sinatra—invisible, unheard voices, passing by at the speed of light. I can turn them into sound with a little box of electronics. And pictures, too. Images from the Olympics, presidential campaigns, foreign wars, natural calamities. The air quavers with visual images borne on oscillations of incredible swiftness. No, it is not the air that quavers, for these invisible radiations leave the atmosphere of Earth and pass without diminishment into the vacuum of space. It is space itself that quavers, at frequencies of millions of oscillations a second. All space—this house, this room, the cavities of my heart—magically, miraculously tremulous with sights and sounds encoded as modulations of electromagnetic waves.

Electromagnetic waves were predicted theoretically by the Scottish physicist James Clerk Maxwell in 1864, an invention of sublime fabulousness. They did not become part of scientific reality until twenty-two years later, when they were experimentally demonstrated by the German physicist Heinrich Hertz, who in effect made the first radio broadcast and reception. At Hertz's transmitter a spark jumped back and forth between two metal spheres fifty million times a second. Across the room a similar spark was instantly produced at the receiver. Invisible, unseen electrical energy had passed through space at the speed of light. It was an unpretentious beginning for the age of radio and television: a tiny spark dancing between two spheres. I visited an exhibit of Hertz's experimental apparatus at the MIT Museum, in Cambridge, Massachusetts. (Actually, the exhibit was of replicas of Hertz's apparatus, built in the late 1920s by German model-maker Julius Orth, working from the originals). The first radio transmitter and receiver have a basement-workshop simplicity about them. Hertz proved the existence of Maxwell's fabulous waves with constructions of wood, wire, string, and sealing wax. The replicas of his apparatus are beautiful in an antique sort of way—yellow varnish, tarnished metal, the patina of crusty wax. They take us back to the day when a clever person with a halfway-decent workshop and a knack for construction could unravel grand mysteries of the cosmos.

In recognition of his achievement, Hertz's name has been adopted as the international unit for frequency; a Hertz is an oscillation of one cycle per second. As I write, Robert J. Lurtsema's "Morning Pro Musica" enters my room riding a wave that oscillates at 89.7 megahertz. From a roomful of invisible oscillations at hundreds of different frequencies my little receiver picks out just the right wave and converts its modulations into sound. At the moment, it's a Bach harpsichord concerto, passing by at the speed of light, a dragon or hippogriff caught on the wing. Fabulous. And true.

It is not the degree of "fabulousness" that distinguishes pseudoscience from science. What makes the fabulous inventions of science "science" is that they can be tested against experience in a communal, reproducible way—as Hertz verified the existence of Maxwell's imagined waves. Yet pseudoscience flourishes, while interest in science languishes. Several recent surveys suggest that America is a nation of science illiterates. The Public Opinion Laboratory of Northern Illinois University polled 2041 adult Americans for the National Science Foundation. More than half of those surveyed did not know that the earth goes around the sun in one year. Twenty percent believed the sun orbits the earth. Another 17 percent thought the earth goes around the sun once a day. Asked whether electrons were smaller than atoms, less than half said yes. Twenty percent said electrons were larger than atoms and 37 percent did not know. One out of five Americans believes sound travels faster than light. In another poll conducted by the National Science Board, nearly half of respondents disagreed with the statement that humans evolved from earlier forms of life. About the same proportion believe that rocket launches affect the weather and that certain numbers are lucky for some people. This is all pretty basic stuff, and indicates not only a failure to assimilate scientific information, but also a certain disconnectedness from the philosophical underpinnings of modern civilization. Undoubtedly, every science educator has a pet theory to explain why so many Americans are ignorant of basic science while at the same time they accept uncritically the goofiest claims for the paranormal. As one who makes his living communicating science—as a teacher and a writer—let me add my two cents' worth to the debate.

When my kids were young they were lucky to have a year of schooling in London, England, and another year in Ireland. An important part of the curriculum in both schools was drawing from nature. Teachers took the students to the park or seashore to sketch what they found: bugs, leaves, blades of grass, shells, stones. The emphasis was not on art, but on observation; not on self-expression, but on faithful representation. The children were asked to look, see, and record what they saw. They were encouraged to honor their senses. In London we lived near the British Museum of Natural History, a vast Victorian storehouse of natural diversity: stuffed animals by the thousands, glass cases full of glistening beetles, iridescent hummingbirds, and gaudy butterflies, room after room of dinosaur bones, rocks, gems, and fossils. The teachers urged the children to go there on weekends. For the deposit of a large brown English penny the kids were given a folding canvas stool, a drawing board, paper, and a fistful of colored pencils. Off they went into the depths of that cavernous building to sketch platypuses and pterodactyls.

Not once in American schools were my children asked to draw from nature. They had art classes, yes, and good ones. They sketched sneakers, bottles, and bowls of bananas—still lifes in the classroom. In biology lab they drew what they saw under the dissecting microscope. But no one ever took them into a natural environment with pencil and paper. They were never asked to sit and sketch a mushroom in the woods or a seashell on the shore. It is my impression that the British and Irish emphasis on drawing from nature had two objectives: developing the child's powers of observation, and reinforcing the child's curiosity about the natural world *as revealed by the senses*. Curiosity and observation are essential foundations for the study of science.

The child who has nightly watched the motions of the stars and planets will find it easier to appreciate that the earth goes around the sun. The child who has observed the clouds, their heapings and tumblings, their dark massings and silver linings, will be better prepared to understand the relationship between rocket launches and weather. The child who has considered the beauty of a heron rising from the pond or the cunning of the spider's web will be less reluctant to acknowledge our evolutionary relationship with other animals. And the child who has paid close attention to the threads of physical causality that stitch nature into a harmonious whole will be properly skeptical of lucky numbers, astrology, and UFOs.

Perhaps the reason we, as a people, are disconnected from science is that we are disconnected from the natural world that science describes. Science is not just a body of information. And science is more than a method. Yet these are frequently the only things that are taught in the schools. Scientific information and scientific method are important and must be taught, especially in the upper grades. But more fundamentally, science is a set of attitudes about the world. Science is the conviction that the world is ruled by something more than chance and the whims of gods. Science is confidence that the human mind can make some sense of nature's complexity, and, almost paradoxically, science is humility in the face of nature's complexity. Science is respect for the evidence of the senses—seeing things *as they are,* and not as we wish them to be. And science is the courage and self-confidence to accept nature's indifference to our personal predicaments. Until young people have assimilated these attitudes, they will be distrustful of scientific information and skeptical of scientific method, and we will continue to be a nation of superstitious science illiterates. These attitudes are not things a teacher can teach. But they are taught by nature. And that's why I am grateful to the British and Irish teachers who sent my kids out into nature with sketch pad and pencil—to carefully observe, meticulously describe, and learn.

CHAPTER 19

The Ink of Night

A *New Yorker* cover by Eugene Mihaesco is tacked on the wall above my desk. The drawing on the cover is simple. A pen lies on a white table, its nib dark with ink. An ink bottle stands open. The ink in the bottle is a map of constellations of the northern sky—Ursa Major, Ursa Minor, and Draco—including the stars Dubhe, Merak, and Mizar. Simple, yet hauntingly provocative. Again and again I pause in my work to look at the drawing. It seems to suggest that the possessor of the pen (Who? A poet? An artist? An astronomer?) draws inspiration from the ink of night. But the star map in the bottle, with its thin white lines defining constellations and star names, is itself a work of creative imagination. So the ink is both night and an image of night. Ambiguous? Certainly. Precisely this ambiguity lies at the heart of science.

Does science describe reality, or does science invent reality? The question is as old as Parmenides and shows no sign of resolution. I suspect that most scientists are willing to live with the ambiguity. They are confident that their theories describe reality, but they also know that theories are creations of the human mind. Consider that little patch of night in the ink bottle, the constellations near the northern pole. Few star groups have so excited the human imagination as Ursa Major, the Great Bear. The seven brightest stars of the constellation—the stars we know as the Big Dipper—are so instantly recognizable that sometimes I wonder if the pattern might not be genetically encoded in the human brain, the way some birds are endowed with an ability to navigate by stars. Those seven gorgeous presences demand recognition. The Greeks provided a charming legend. The nymph Callisto was loved by Zeus, who transformed her into a bear to protect her from the wrath of Hera, his jealous spouse. One day Callisto's son Arcas was out hunting in the forest and raised his bow to shoot the bear, not recognizing his mother in altered form. Zeus observed the impending tragedy from Mount Olympus, and speedily intervened. He changed Arcas into a little bear, and placed mother and son into the heavens where they remain today, the Great Bear and the Little Bear, arched poignantly toward each other, eternal victims of Zeus' wandering eye. But Hera had the last laugh; she moved the two bears into the part of the sky near the celestial pole, where they never set and therefore never rest.

There was a time when images of bears and the story of Callisto and Arcas might have satisfied our curiosity about the polar stars, but the tension between experience and story has become too slack for the story to any longer have currency as science. Today we require new stories, stories more closely tied to our experience of the stars and more consistent with our other knowledge of the world. Let me dip my pen into Mihaesco's ink of night and tell the astronomer's story of Dubhe, Merak, and Mizar, the three stars that are named on the map in the bottle. Dubhe, the star at the lip of the Dipper, is a yellow-orange giant star ten times larger than the sun and a hundred times more luminous. It lies one hundred light years from Earth, a distance so vast that it would take a Voyager spacecraft, such as the craft we sent to Jupiter, Saturn, Uranus, and Neptune, a million years to get there. Dubhe was once a star very much like the sun, but it has depleted its energy resources and entered its death throes, swelling up to devour its inner planets and boiling away whatever oceans and atmospheres those planets might have had. Dubhe's fate will someday be the fate of our sun. Merak and Mizar, at the bottom front of the Dipper's bowl and at the bend of the handle respectively, are sibling stars, born from the same great gassy nebula and streaming together through space from the place of their birth. They are stars in the prime of life, many times brighter than the sun, and almost certainly accompanied in their travels by families of planets. Mizar is a wonderful thing to behold through a telescope. It is actually a system of two great suns, bound together by gravity, circling about a common center of attraction once every ten thousand years.

How is it that astronomers can tell such stories, stories more fabulous than any myth of gods and nymphs, when the ink of night offers to the eye only pinpricks of light? The answer is both simple and complex. *We look, we invent, we look again.* We test our inventions against what we see, and we insist that our inventions be consistent with one another, that our stories of the stars be consistent with our stories of the earth, of life, and of matter and energy. And tension! Always we are testing the tension of the instrument that is science, observing that the strings of theory are taut and resonant, and that every ear hears an identical note when the strings are bowed. The same theories of gravity and dynamics describe the fall of an apple from a tree and the streaming of stars through space. The story of the falling apple and the story of the stars must resonate together. Only then, when our stories of the world vibrate with a symphonic harmony, are we confident that our inventions partake of reality.

The starry night is both the ink of invention *and* the invention; it is an ambiguity we have learned to live with. We are confident (or we wouldn't do science at all) that out there in space, six hundred trillion miles from Earth, the dying star Dubhe burns with the brilliance of a hundred suns, and Merak and Mizar stream together from a common point of origin. But we also know that the astronomer's stars, like the stars of the Greek storytellers, are constructed of the frail stuff of likeness. Even the sturdiest of scientific concepts is rooted in metaphor.

Reaching for a book on a high shelf I tipped down *Season Songs* by poet Ted Hughes, which attracted attention to itself by delivering a lump on the head. I sat on the floor and read again those nature poems written twenty years ago by Britain's present poet laureate.

> *Fifteenth of May. Cherry blossoms. The swifts*
> *Materialize at the tip of a long scream*
> *Of needle—"Look! They're back! Look!"*

At Hughes's invitation I watched the swifts, those quickest of birds, watched their "too-much power, their arrow-thwack into the eaves." Arrow-thwack! Yes, I thought, that's exactly right. That's exactly the way swifts zip into the eaves of the old barn on their evening high-speed revels. As if shot from a bow. Too quick to be animate. Then, at poem's end, a lifeless young swift is cupped in the poet's hand in "balsa death." What a phrase! By it we are made to feel the surprising unheaviness of the bird in the hand, the hollowed-out bones and wire-thin struts beneath the skin of feathers, a tiny machine perfected by one hundred million years of evolution to skim on air, as light as balsa. Hughes's delightful images reminded me how much scientists need poets to teach us how to see.

Scientists are trained in a very un-metaphorical way of seeing. We are taught to look for *immediate* connections: X causes Y, Y causes Z. We strip away the superfluous, the noncausal. We isolate. We weigh and measure. The average density of a bird is significantly less than the average density of a mammal of comparable size. That's one reason birds fly. Balsa wood has nothing to do with it. But anyone who has held a bird in the hand will recognize the aptness of Hughes's balsa wood image, the curious absence of expected heft. The metaphor is instructive. We learn something about birds that no ornithological text quite so vividly conveys. Make no mistake, I am not dismissing the scientific way of seeing. Weighing, measuring, abstraction, and dissection have proved their worth as royal roads to truth. But the poet's eye guides us to truths of another kind. No field biologist has seen "hares hobbling on their square wheels," yet Ted Hughes's metaphor is so perfectly truthful we can't help but laugh. No ichthyologist has recorded the mackerel's "stub scissors head," but we readily imagine the blunt jaws of the fish shearing open and shut as if operated by a child's deliberate hand. No astronomer has watched a full moon that "sinks upward/ To lie at the bottom of the sky, like a gold doubloon," but Hughes's lunar image truthfully reminds us that there's no up or down to the bowl of night.

Philosophers of science insist that science is metaphorical. They cite, for example, Newton's "clockwork" solar system and Robert Boyle's elastic "spring" of air. Christian Huygens, a Dutchman who lived by water, first thought of light as a "wave." Alfred Wegener, a meteorologist who traveled in the frozen arctic, conceived of continents drifting like "rafts" of ice. The philosophers are right: At root, all scientific knowledge is metaphorical. The stars of the astronomers (colossal suns, hugely distant fires, furiously blazing globes of gas) are as unlike the objects of actual experience (pinpricks of cold light in the dome of night) as are the nymphs and randy gods of Greek myth—and no less metaphorical. But young scientists are not trained to think (or to see) metaphorically and we may be poorer for it. Metaphor is a way of seeing noncausal connections, as when Ted Hughes speaks of April "struggling in soft excitements/ Like a woman hurrying into her silks." On the face of it, there's nothing in the metaphor of use to a scientific student of the seasons, yet the words significantly alter our perception of spring. "Struggle," "soft," "excite," "hurry," and "silk" force us to think about spring in layers and levels of meaning.

Scientists, especially those working in narrow areas of specialization, are often trapped by tunnel vision. Metaphors have a way of exploding the bounds of perception, of making plain the essential unity of nature. Great science often happens when likenesses are perceived where none were thought to exist: Life is a "tree." The electron is a "wave." Thermodynamic systems are "information." In his best-selling book *Chaos: Making a New Science,* James Gleick describes how people working in widely different areas of science came to understand that certain apparently diverse phenomena had much in common. A dripping faucet, a rising column of cigarette smoke, a flag flapping in the wind, traffic on an expressway, the weather, the shape of a shoreline, fluctuations in animal populations, and the price of cotton: All these things, it turns out, can be described by a new kind of mathematics—fractal geometry and its variations—based on randomness and feedback. The new chaos scientists, says Gleick, are reversing the reductionist trend toward explaining systems in terms of their constituent parts, and instead are looking at the behavior of whole systems. Their ability to see likenesses between systems is the key to their success.

And that's what poets can teach scientists. Perhaps a course in metaphor should be as important a part of a scientist's training as a course in mathematics. When Ted Hughes writes . . .

The chestnut splits its padded cell.
It opens an African eye.
A cabinet-maker, an old master
In the root of things, has done it again.

. . . he may be on to more than he knows. The old master at the root of things is metaphor.

Science and poetry are both metaphorical, but science is not poetry. Robert Frost once said that writing free verse is like playing tennis with the net down; it is not a sentiment I happen to agree with, but if Frost's simile is accepted, then doing science is like playing tennis on a court thick with nets and white lines, constrained by hugely complicated rules. Profound strictures on the language of science are imposed by the requirement that science be public knowledge, internally consistent, reproducible, and (in its expression) as unambiguous as possible. Metaphor may be the spark that ignites scientific understanding, but the expression of the flame soon leads into a fractured maze of specialized vocabularies.

In his *Field Guide to the Birds,* Roger Tory Peterson gives this characterization of the purple finch: "Male: About the size of a House Sparrow, rosy-red, brightest on head and rump." And then he adds a traditional description: "a sparrow dipped in raspberry juice." That's it. Decisive. The perfect fit. Anyone who has ever seen a purple finch will recognize the aptness of that final phrase. The raspberry juice image is perfect—for the poet or amateur birdwatcher—but it hardly qualifies as science. Science has as its task elucidation of the real, unarbitrary connections between things, and "sparrow dipped in raspberry juice" isn't terribly helpful. Sparrows and finches do belong to the same family (Fringillidae) of the perching birds (order Passeriformes, class Aves), but they are different enough in subtle anatomical ways to be classified in separate genuses. The genus and species designations of the purple finch (*Carpodacus purpureus*) tell us more about the bird's proper place in the tree of life than Peterson's evocative metaphor.

Popular speech often muddies the water of understanding. Consider these examples from botany: The asparagus fern that grows at our kitchen window is not a fern at all. It is a seed-bearing plant of the lily family, although, superficially, it certainly looks more like a fern than a lily. The strawberry begonia, also called strawberry geranium, is not a strawberry, nor a begonia, nor a geranium, and is not in any of those plant families. It is instead *Saxifraga,* a genus closely related to the roses. It is the technical designations of the plants, not their popular names, that convey reality to the botanist. And here, in the forced retreat into specialized vocabularies, science begins to take on a cold and distant aura, and here also many people desert science for the easily accessible, anthropocentric metaphors of pseudoscience.

Ordinary language is so steeped in misconceptions that scientists often find it best to start from scratch, inventing purpose-made languages. In the introduction to his *Wildflower Guide,* Peterson lists sixty ways that a botanist can say that a plant is not smooth: aculeate, aculeolate, asperous, bristly, bullate, canescent, chaffy, ciliate, ciliolate, coriaceous, corrugated, downy, echinate, floccose, flocculent, glandular, glanduliferous, glumaceous, glutinous, hairy, hispid, hispidulous, and so on. The dictionary defines both "aculeate" and "echinate" as prickly, and one might reasonably ask why a good English word like prickly won't serve the purpose. The botanist, I am sure, has an answer, presumably involving subtle shades of difference. Those shades of difference are the basis of the botanist's superior knowledge of plants. We can only know something if we can say it, so it is perhaps inevitable that the language of science becomes more specialized as we understand the world in ever greater detail. The Eskimos have a dozen words for snow, and in the Arabic language there are thousands of words associated with camels. If you live all of your life on snow or with camels, then a dozen, or even a thousand words, are barely sufficient to describe your experience. And if your life is devoted to the study of plants, then a stem that is aculeate or echinate may appear significantly different from one that is merely prickly.

Some linguists insist that the language we speak actually determines what we see. This remarkable idea was first suggested by the nineteenth-century German philologist Wilhelm von Humboldt, who said, "Man lives with the world about him, principally, indeed exclusively, as language presents it." In other words, if you have a word for it, then it exists; if you don't have a word for it, it doesn't exist. This idea was taken up in our own century by the linguist Edward Sapir and his student Benjamin Lee Whorf, and has come to be known as the Sapir-Whorf hypothesis. According to this school of thought, the vocabulary and grammatical structure of a language place powerful constraints on how the speaker perceives the world. Whorf's own work was primarily with the language of the Hopi Indians of the American Southwest; he offered many examples of how Hopi language and Hopi science go hand in hand, and how they differ in significant ways from European languages and science.

Of course, the problem with the Sapir-Whorf hypothesis is the problem of the chicken and the egg. Does language determine experience, or does experience determine language? (Which is another way of expressing the question which began this essay: Does science describe reality, or does science invent reality?) In recent years the Sapir-Whorf hypothesis has gone rather out of fashion; nevertheless, the two linguists and their followers clearly demonstrated a close connection between language and perception. All of which helps explain why scientists find the need to invent specialized vocabularies. Words like bullate, glanduliferous, and hispid do not trip lightly from the tongue, but they are well suited for their purpose. Each word gives precise expression to something seen, and helps free perception from the muddy imprecisions of popular speech.

Philosopher of science John Ziman says this of scientific communication: "Vivid phrases and literary elegances are frowned upon; they smell of bogus rhetoric, or an appeal to the emotions rather than to reason. Public knowledge can make its way in the world in sober, puritan garb; it needs no peacock feathers to cut a dash in." It is easy, even for the scientist, to regret the brusque inelegance of specialized scientific languages, and long for the graceful poet's metaphor ("struggling in soft excitements"), but that is the price we pay for clarity of thought. Truth is not served by making the world conform to the soft contours of our tongue. Language must be made to serve reality, rather than the other way around. Calling the asparagus fern a lily ("Perianth usually conspicuous, not chaffy, regular or nearly so, 6-parted; stamens hypogynous or adnate to the perianth; pistil 1; ovary 3-celled, usually superior") may offend common sense, but it makes perfect sense as science. And calling the purple finch "a sparrow dipped in raspberry juice" may be deliciously poetic, but—scientifically speaking—*Carpodacus purpureus* is the bird's real name.

CHAPTER 20

A Measure of Restraint

On September 13, 1987, two unemployed young men in search of a fast buck entered a partly demolished radiation clinic in Goiânia, Brazil. They removed a derelict cancer therapy machine containing a stainless steel cylinder, about the size of a gallon paint can, which they sold to a junk dealer for twenty-five dollars. Inside the cylinder was a cake of crumbly powder that emitted a mysterious blue light. The dealer took the seemingly magical material home and distributed it to his family and friends. His six-year-old niece rubbed the glowing dust on her body. One might imagine that she danced, eerily glowing in the sultry darkness of the tropic night like an enchanted elfin sprite. The dust was cesium-137, a highly radioactive substance. The lovely light was the result of the decay of the cesium atoms. Another product of the decay was a flux of invisible particles with the power to damage living cells. The girl is dead. Others died or became grievously sick. More than two hundred people were contaminated.

A beautiful, refulgent dust, stolen from an instrument of healing, had become the instrument of death. The junk dealer's niece was not the only child who rubbed the cesium on her body like carnival glitter, and the image of those luminous children will not go away. Their story is a moral fable for our times—a haunting story, touched with dreamlike beauty and ending in death. It evokes another story that took place almost a century ago, another story that illustrates the risks that are sometimes imposed by knowledge. It is a story of Marie and Pierre Curie, the discoverers of radium, as told by their daughter Eve.

The story begins at nine o'clock in the evening at the Curies' house in Paris. Marie is sitting at the bedside of her four-year-old daughter, Irene. It is a nightly ritual; the child is uncomfortable without her mother's presence. Marie sits quietly near the girl until the restless young voice gives way to sleep. Then she goes downstairs to her husband Pierre. Husband and wife have just completed an arduous four-year effort to isolate from tons of raw ore the tiny amount of the new element that will win them fame. The work is still on their minds: the laboratory, the workbenches, the flasks and vials. "Suppose we go down there for a moment," suggests Marie. They walk through the night to the laboratory and let themselves in. "Don't light the lamps," says Marie, in darkness. Before their recent success in isolating a significant amount of the new element, Pierre had expressed the wish that radium would have "a beautiful color." Now it is clear that the reality is better than the wish. Unlike any other element, radium is spontaneously luminous! On the shelves in the dark laboratory precious particles of radium in their tiny glass receivers glow with an eerie blue light. "Look! Look!" says Marie. She sits down in darkness, her face turned toward the glowing vials. *Radium. Their radium!* Pierre stands at her side. Her body leans forward, her eyes attentive; she adopts the posture that had been hers an hour earlier at the bedside of her child. Eve Curie called it "the evening of the glowworms."

Marie and Pierre Curie and their new element became famous. By the middle of the first decade of this century had begun what can only be called a radium craze. A thousand and one uses were proposed for the material with the mysterious emanations. The curative powers of a radium solution—called "liquid sunshine"—were widely touted. It was soon discovered that radium killed bacteria, and suggested uses included mouthwashes and toothpastes. Health spas with traces of radium in the water became popular. Entertainers created "radium dances," in which props and costumes coated with fluorescent salts of radium glowed in the dark. It is said that in New York people played "radium roulette," with a glowing wheel and ball, and refreshed themselves with luminescent cocktails of radium-spiked liquid. The most important commercial application of radium was in the manufacture of self-luminous paint, widely used for the numerals of watches and clocks that could be read in the dark. Hundreds of women were employed applying the luminous compound to the dials. It was a common practice for them to sharpen the tips of their brushes with their lips. Many of these women were later affected by anemia and lesions of the jawbone and mouth; a number of them died.

By 1930 the physiological hazards of radioactivity were recognized by the medical profession and the reckless misuse of radium had mostly ceased. But the mysterious emanations—which properly used are an effective treatment for cancer—had taken their toll. Marie Curie discovered the secret of the stars; her tiny glass vials contained the distilled essence of the force that makes the universe glow with light. She died of radiation-induced leukemia, with cataracts on her eyes and her fingertips marked by sores that would not heal. Like many of the gifts of knowledge, radium had proved a mixed blessing. The poet Adrienne Rich has described Marie Curie's death this way:

> *She died   a famous woman   denying*
> *her wounds*
> *denying*
> *her wounds   came   from the same source as her power*

The evening of the glowworms! Eve Curie's evocative phase might also be used to describe the dance of the Brazilian children, their bodies luminous with cesium-137. In these two stories we are drawn at last and emphatically into the circle of the Janus-faced god. Death and beauty, wounds and power: The piercing horns of the dilemma of science, demanding from the seeker of truth a measure of restraint.

As I write these lines, I recall glowworm evenings I experienced as a child in Tennessee, running barefoot with my young companions through the lush green grass of the long sloping lawn, catching up fireflies in our hands. Stars in the silky night glimmered in concert with insect scintillations—tiny flashes of cold brilliance reflected in a canopy of over-arching pines, as in dark water. The insect lights seemed a miracle, a conjuration of elfin magic; a dozen fireflies in a bottle made a fairy light. Now, forty-five years later, I have before me as I write the image of another firefly light: a photograph of a tobacco plant made to glow with the phantasmic radiance of the firefly's luciferous gene. I have clipped the photograph from the pages of the journal *Science* and tacked it up on the wall above my desk. It expresses what is best and worst in our quest for knowledge.

To make the autoluminescent tobacco plant, genetic engineers first located the firefly gene—the DNA segment that gives rise to the enzyme that catalyzes the firefly's light-producing chemistry. The purloined gene was then introduced into the cells of tobacco plants, and the plants watered with a solution of the chemicals necessary for the luminescent reaction. The plants then emitted a faint but detectable light. The photograph was made by placing a genetically altered plant in contact with photographic film for twenty-four hours. The result is a scientific artifact that qualifies as a work of art.

One hardly knows how to react to experiments such as this.

One admires the knowledge and skill that enabled the genetic researchers to achieve so remarkable a transmutation of living matter—a plant made luminous with an animal's gene. Certainly, one is moved to a deeper respect for the chemical machinery of life. Still, I turn to the photograph of the genetically altered plant with a sense of foreboding. The tobacco plant seems to rise out of the paper like a will-o'-the-wisp or friar's lantern, those flickering phosphorescent lights that are sometimes seen over marshes and swamps at night, that in folk legend beckon unwary followers into the mire.

The transgenic researchers do not consider their experiments frivolous or dangerous. They are confident that the firefly's etheric gene can be spliced with other genes as a valuable marker in genetic experiments. Researchers need to know quickly if and where transplanted genes have been activated. The firefly's light, issuing from the cells of another organism—human cancer-fighting cells, for example—can be an ideal signal. There is no doubt that the tobacco-cum-firefly experiments, and others like them, will lead to discoveries of potential benefit to society: Grains that are resistant to disease, fruit trees that defy frost, bacteria that eat oil spills, vaccines for the cure of animal and human diseases—all these things and more are promised by genetic engineers. Then what is the source of my uneasiness? Certainly, genetic engineering is not the first breakthrough in science that harbored potential for danger as well as good: The discovery of radium comes too quickly to mind. Radium beckoned us forward with the promise of cures for disease and inexhaustible energy resources. Then Janus turned to reveal his other face—terrible weapons of destruction, a plague of nuclear waste, cancers caused, not cured. In many ways, the fruitful promise of genetic engineering is greater than that of radioactivity, but so is the potential danger. A gene reproduces. A gene copies itself into the fabric of life. Nuclear waste remains radioactive for thousands of years; a gene is potentially immortal. The soft phosphorescent light of the genetically altered tobacco plant beckons us toward a future bright with health and plenty, but it also has a spooky Frankensteinian quality that warns us to proceed with caution. "For Beauty's nothing but the beginning of Terror," wrote the poet Rilke, and all too often his words might describe the enterprise of science.

On those sultry nights in Tennessee we caught glowworms in our hands. Sometimes we pinched their tiny bodies to set their gene-activated fires alight. But we squeezed gently, and then released the insects to take their place again among the live constellations of the summer night. We recognized, if only in a childlike way, an integrity and balance within nature that demands of earth's dominant species a judicious self-restraint. The unexamined quest for knowledge is hemmed with peril.

"Scientific curiosity is not an unbounded good." One does not often hear those words, especially uttered by a scientist. I quote them from an essay addressed to scientists by the octogenarian biochemist Erwin Chargaff. Chargaff's cautionary comment was prompted by developments in genetics, molecular biology, and embryology, and particularly in the technology of human reproduction. In effect, he charges researchers with knowing *too much* about the molecular machinery of life, and with using that knowledge to "stick our fingers into the incredibly fine web of human fate." Research on human embryos especially arouses Chargaff's disapproval. He fiercely condemns *in vitro* fertilization, the freezing of embryos for later implantation into a mother's womb, surrogate motherhood (especially for a fee), and various forms of transgenic tinkering.

Chargaff dismisses as so much quibbling the question of when an embryo becomes "human"; the life of the embryo begins, he believes, with the fertilized egg, and deserves the same respect from researchers as any other human life. He takes note of the human benefits that are put forward to justify embryonic research—the correction of genetic defects, helping childless couples have children, and so forth. But with lofty defiance, he dismisses the idea that the end might justify the means. Even more disturbing, Chargaff suggests that the proffered "justifications" for embryonic research sometimes mask the real motives—the avarice and ambition of researchers. It is a serious charge, and one that in my view is largely unjustified. It is a charge that raises the hackles, even the anger, of those involved in reproductive research. But Chargaff's stern challenge carries the weight of a fruitful life in science. He is emeritus professor of biochemistry at Columbia University and is best known for his demonstration in the late 1940s that certain chemical components of DNA molecules always occur in constant ratios, a result that was crucial to the discovery of the structure of the DNA double helix by Watson and Crick. Chargaff was among the first to recognize that the chemical composition of DNA was species specific. He has won many international awards for a lifetime of pioneering work in biochemistry.

Chargaff has made himself something of a career as a scientific gadfly. His prescriptive moralizing is dismissed by many scientists as sour sentimentality and antiprogressive romanticism, but his words fall upon other ears with a kind of Jehovian thunder—and rightly so. Whether Chargaff's castigation of contemporary embryonic research is philosophically or morally correct is debatable; but that science should value its Chargaffs—thundering from on high—is beyond dispute. Sometimes is it necessary for the grand old men of science, no longer caught up with the self-serving activities of making a career, to question the moral implications of what we do. In setting himself up as the judge of science, Chargaff wins few plaudits; there are no Nobel prizes for curmudgeons. But as Chargaff himself once wrote, "Philosophy is one of the hazards of old age."

Erwin Chargaff spent his childhood in Austria, in what seemed to him the last golden rays of a more civilized era. He was watching the younger sons of Kaiser Wilhelm II play tennis when news came of the assassination of the Austrian archduke Franz Ferdinand, an event that plunged all of Europe into darkness. The years between the wars were spent in Vienna, where Chargaff took his degrees. Torn between science and the study of literature, he drifted into chemistry, as later he drifted into biochemistry. He was forced to leave Europe by the rise of the Nazis. Again darkness descended. His mother was deported from Vienna into oblivion. In his autobiography, Chargaff says of his life: "In the Sistine Chapel, where Michelangelo depicts the creation of man, God's finger and that of Adam are separated by a short space. That distance I called eternity; and there, I felt, I was sent to travel." He has been at every moment of his life aware of the immensity of the darkness that is nature. As a scientist, he lessened the darkness with the light of his own genius. Now, as a respected professor emeritus surrounded by solved riddles, he remains struck by how *little* we understand—and made anxious by how *much* we understand. The darkness of ignorance and the light of knowledge equally seem to frighten him. In certain contemporary research, Chargaff apparently feels that science comes dangerously close to bridging the gap between God's finger and the finger of man. In asking us to hold back he gives voice to widely popular concerns, and throws into sharp relief our unsettling ambivalence toward science. "Restraint in asking necessary questions," he wrote, "is one of the sacrifices that even the scientist ought to be willing to make to human dignity."

Self-luminous children dancing in the tropic night, the glowworm light of radium issuing from glass vials in a darkened lab, tobacco plants chandeliered with firefly light—all friar's lanterns beckoning us to bridge the gap between God's finger and the finger of man. Many scientists hold that knowledge is an absolute good, to be pursued at any risk. Light is invariably better than darkness, they say; we must not forget that the glowworm light of the

Brazilian children was stolen from an instrument of healing, and the glowworm light of the transgenic tobacco plant illuminates the darkness of ignorance and superstition.

Erwin Chargaff's challenge disturbs the sleep of reason. What he says about embryonic research applies with equal (perhaps greater) force to transgenic experimentation, nuclear research, and many other areas of contemporary science: The unshuttered light of knowledge threatens moral conflagration, and the unconstrained exploitation of nature holds the potential for shattering annihilation; the source of our power is also the source of our wounds. Chargaff believes humans cannot live without mysteries, and yet he has devoted his life to unraveling the greatest mystery of all, the mystery of human life. He contributed mightily to discovering the secret of DNA, and yet damns the use to which that knowledge has been put. He is a man of reason who agrees with Goya that "the dream of reason brings forth monsters." Struck by these apparent inconsistencies between Chargaff's life and words, many scientists dismiss his critique of contemporary research as cantankerous obfuscation. Unbounded scientific curiosity, they say, has proved its worth, and whatever are the dangers implicit in knowledge, they are far outweighed by good. Indeed, they say, in turning his back on contemporary research, Chargaff would have us return to a time when human life was the helpless plaything of poverty, disease, and death. Chargaff answers in a voice honed to a fine sharp edge on the whetstone of paradox: "A balance that does not tremble cannot weigh. A man who does not tremble cannot live."

CHAPTER 21

The Virgin and the Mousetrap

In the year 1303, Enrico Scrovegni, a businessman of Padua, Italy, commissioned the construction of a chapel, known today as the Arena Chapel, partly to expiate the sins of his father, a notorious moneylender assigned by the poet Dante to the seventh circle of Hell. The building is decorated with frescoes by the artist Giotto di Bondone, showing scenes from the lives of Mary and Christ. One of these, the *Adoration of the Magi*, is famous for the special nature of its Christmas star. The celestial object that hovers above the stable at Bethlehem is not the traditional many-pointed star with rays streaming down toward the Christ child, but a remarkably realistic comet. Its "rays," or tail, point upward into the evening sky, like the tail of a real comet streaming away from the setting sun. Roberta Olson, an art historian at Wheaton College in Norton, Massachusetts, has argued convincingly that the "star" in Giotto's painting is Comet Halley. Guy Ottewell and Fred Schaff, authors of *Mankind's Comet*, a comprehensive historical and astronomical survey of Comet Halley, agree that the famous comet was Giotto's inspiration.

Comet Halley appeared in Italian skies in the year 1301, only a few years before Giotto began work on the Paduan frescoes. At that time the painter was probably in Florence, and may have observed the comet in the company of his friend, the poet Dante, who also resided in that city. With the help of a star globe and computer, I have reconstructed what Giotto and Dante might have seen. The comet was brightest in late September, appearing low in the northwest after sunset, with tail streaming upward toward the Big Dipper. The posture of the comet in Giotto's painting is strikingly similar to that of the real comet in the sky, although reversed left to right by the artist, presumably for compositional reasons. Further, the coma and tail of Giotto's comet are very like photographs of Comet Halley made during its 1910 apparition.

Giotto's use of Halley as the Christmas star was not farfetched. Comets had long been associated with the births of kings and commencements of new dynasties, and several early Christian theologians had assumed that a comet prefigured the birth of Christ. In Giotto's time, the periodic nature of comets was not yet recognized, and the mysteriously beautiful object hanging low in the western sky must have seemed an appropriately miraculous model for the Christmas star.

Giotto is best known in the history of art as a pioneer of naturalism. His figures have mass and volume, and relate to one another in postures that are true to nature. The paintings are not mere icons, serving a religious or ritualistic purpose as did the Gothic cathedrals dedicated to the Virgin. Instead, the viewer feels himself a part of the scene which the artist has created. It has been said that Giotto elevated painting from the service of symbolism and made it the mirror of mankind. More significantly, Giotto's paintings are mirrors of the world. In *Adoration of the Magi* the artist proves himself a careful observer of the sky, and gives us the first-ever rendering of a comet that is almost photographically accurate. His emphasis on exact observation and realistic representation reflects the end of the Middle Ages and the beginning of the Renaissance.

Padua, the city where Giotto so beautifully rendered Comet

Halley, was three centuries later the place where Galileo made his own historic observations of the sky. Giotto and his successors, the artists of the early Renaissance, pioneered new concepts of space, time, and mass that would come to fruition in the physics of Galileo and Newton. Giotto's observations of Comet Halley, presumably made during September, 1301, were direct historical antecedents for the celestial observations of Copernicus, Tycho Brahe, and Kepler. Renaissance artists honored the senses as the only reliable arbiters of truth. They confidently asserted the capacity of human reason to make some sense of the world, and they passionately believed in the value of unfettered curiosity. These are the very qualities of thought that animate the scientific enterprise. Western science had its beginnings in Western art, and perhaps nowhere more importantly than in the work of Giotto di Bondone.

On March 13, 1986, only a few weeks before I made my pilgrimage to Australia in pursuit of Comet Halley, a space probe launched by the European Space Agency passed within four hundred miles of the comet's nucleus. It was the ninth visit of Comet Halley to the inner solar system since the apparition observed by Giotto, and the twenty-sixth since the time of Christ. A camera aboard the spacecraft radioed back to Earth amazingly detailed photographs of the comet's nucleus: a potato-shaped object, ten miles long, marked with hills and valleys and effusing clouds of luminous gas. A spectroscope within the craft detected in the gas familiar terrestrial substances, including the very organic molecules that are the stuff of life. The spacecraft was named *Giotto,* in honor of the pioneering artist who gave us the first naturalistic portrait of Comet Halley.

The encounter of spacecraft *Giotto* with Comet Halley can be taken as a culmination of the enterprise begun by Renaissance artists. For thousands of years, human beings had considered themselves the playthings of fortune, watched over by celestial presences, ruled by stars. Comets, when they appeared, were frightening and unexpected occurrences, announcing the arbitrary intervention of gods into human affairs. Now, through the agency of science, we have reached out and touched the comet, affirmed its thralldom to the rule of law, and found hidden within its effusing atmosphere the organic substances of our own existence. The encounter of the spacecraft with the comet is a supreme expression of Renaissance self-confidence, an apotheosis of *Homo curiosus* into the realm of the deposed gods. Ironically, the visit to the comet also confirmed our cosmic mediocrity and nature's sublime indifference to human affairs. This is the message of spacecraft *Giotto:* We are indeed the detritus of stars, our atoms are the flour of celestial mills, our bodies the bakings of solar fires. We are not ruled by stars; we and the stars are one.

Cosmic mediocrity. Nature's indifference. Our bodies (and our souls) the detritus of stars. These are hard lessons. To accept the scientific vision of reality requires courage and imagination. Yes, mediocrity means that we are not the central miracle of creation; rather, all of nature is miraculous. Yes, indifference means that the sky is deaf to our imprecations; rather, with our intellects and imaginations we reach out to comprehend the sky. Yes, we are the detritus of stars, mere matter; but matter is more than we ever dreamed, a stuff of dazzling subtlety and potentiality. Science is more than a mere accumulation of facts about the world. It is an attitude, taught to us first of all by the artists of the Renaissance, who were curious, self-confident, who honored the senses, who were inventive and forward-looking. It is, if you will, as much an attitude about ourselves as about nature; more precisely, it is a willingness to honor ourselves, with our hopes and dreams, our exhilarations and fears, as typical efflorescences of nature's creative potential.

I have before me on my desk a reproduction of the Merode Altarpiece, a painting on three panels by a fifteenth-century Flemish master believed by many scholars to be Robert Campin (the original hangs in the Cloisters gallery of New York's Metropolitan Museum of Art). The triptych depicts the moment of the Annunciation, when the angel Gabriel announces to the Virgin that she is to become the mother of Christ. It is a warm, marvelous work, rich with cultural meaning. The lefthand panel of the triptych shows a pious couple kneeling in reverence at the door of Mary's house; it is perhaps a portrait of the donor of the altarpiece and his wife. The righthand panel presents Joseph at work in his carpentry shop. In the central panel Gabriel is about to utter the momentous words, while Mary reads, as yet unawares. The year is about 1430, a bit more than a century from the time of Giotto's *Adoration of the Magi*. Gutenberg is beginning his experiments with movable type. Before another century has passed we will have witnessed the technological innovations of Leonardo, the anatomical studies of Vesalius, the revolutionary astronomy of Copernicus. In the Merode Altarpiece we are poised on a cusp of history, between the Middle Ages and modern times, between the world of the cathedrals and the world of the dynamos. The angel announces to the Virgin a message of medieval otherworldliness and detachment; the material elements of the painting anticipate the scientific and technological revolution that is about to overtake Western culture.

The quietly unfolding drama of the Merode Altarpiece is set in a typical fifteenth-century Flemish household. Beyond the open door of the courtyard and the window of Joseph's shop we are provided a glimpse of the busy life of the town, coming alive with commerce and technical innovation from the feudal slumber of the Middle Ages. What is most striking about the painting is the artist's keen eye for *things*, for the mechanical accoutrements of the rising middle class. The textures of wood, metal, cloth, and stone are lovingly rendered. Here is the carefully crafted wrought iron of the candle holders and fire irons, the gleaming brass of the hanging wash basin, and the sharp-edged steel of Joseph's tools. Here are things the well-to-do Flemish burgher would have been proud to have in his household: the fine iron lock, the lacquered wood and metal towel rack, the fine porcelain vase, the splendid carved oak settle. Here are two lovingly protected books, one of them in Mary's hands. The garments of the angel and the Virgin are rich and trimmed with gold. On Joseph's bench is a clever mechanical mousetrap that gives the work its popular title, The Madonna of the Mousetrap. It is a "better" mousetrap, perhaps, that perennial symbol of progress and invention. With this one delightful image the artist has captured the spirit of his time: mechanical, inventive, forward-looking, preoccupied with matter and force. In the century that followed the painting of the Merode Altarpiece, science and technology consolidated a new alliance that led to the Scientific and Industrial Revolutions, and to a new era of health and material well-being for a large part of humankind.

Knowledge cannot be unknown and the gifts of technology are seldom refused, yet many of us are alienated from the scientific instruments of knowledge, frightened by the darker excesses of technology, and nostalgic for a "simpler" time when knowledge was as accessible as a book in the hand and the gifts of technology were as proportionate to the human scale as a humble mousetrap. I turn again to the Merode Altarpiece. I admire the selflessness of the prosperous donor and his wife. I observe the quiet pride of Joseph in his craft. I am stuck by the *appropriateness* of the mousetrap. Above all, I am moved by the serene pleasure the Virgin finds in her book. Behind her, in the porcelain vase, are lilies, the symbol of purity of heart. The altarpiece evokes a harmony of material and spiritual concerns, a confluence of practical knowledge and moral aspirations. In this simple household scene, rendered on a cusp of history, the Flemish master has given us a vision of two worlds in perfect balance.

"A balance that does not tremble cannot weigh": I remember the haunting words of Erwin Chargaff. In these essays I have sought points of contact between the scientific enterprise and perennial human questions: Who am I? Where did I come from? Why am I here? The answers suggested by science are not immediately reassuring: We are not after all the lords and purpose of creation. But to reject science because it does not confirm our cosmic centrality is to reject the one instrument that lets us encompass within our minds the central mysteries of creation—life, intelligence, order, and evolution on the cosmic scale. There is a powerful temptation to retreat from acceptance of our cosmic mediocrity into the consoling superstitions of an earlier age or, alternately, to deify science as our new god. But the often anthropomorphic forms of the older faiths are insufficiently capacious to encompass what we have learned about the world, and science is too tentative an instrument to be deified. Science cannot be a repository for ultimate faith: It *is* a fulcrum upon which we can hope to balance the treasure of our knowledge against the claims of ignorance. It is a knife edge honed on recurring disappointment, a place of tremulous rest. It is not comfortable. It is all we have.